T0384048

Integrated Earthquake Simulation

Integrated earthquake simulation (IES) is a new method for evaluating earthquake hazard and disaster induced in a city and urban area, by carrying out a sequence of numerical simulations of such aspects as earthquake wave propagation, ground motion amplification, structural seismic response, and mass evacuation. This textbook covers the basics of numerical analysis methods of solving wave equations, analyzing structural responses, and developing agent models for mass evaluation, which are implemented in IES. It also explains the recent achievement of enhancing IES with the high the performance computing capability that can take advantage of modern computers and the techniques of automated model construction of various numerical analysis methods. Detailed worked examples of IES for the Tokyo Metropolis Earthquake and the Nankai Trough Earthquake are given, which uses large scale analysis models of actual cities and urban areas. Monte-Carlo simulation that accounts for the effects of uncertainties related to earthquake scenarios and modeling undergrounds and structures on the estimate of earthquake hazard and disaster is presented as an advantage of IES.

Muneo Hori received his B.E. degree from The University of Tokyo, his M.S. degree from Northwestern University in 1985 and his Ph.D. degree from the University of California, San Diego, in 1987. He became a professor at the Earthquake Research Institute at The University of Tokyo, in 2001. He moved from the university to Japan Agency of Marine-Earth Science and Technology as a director general in 2019. His research interests are in the areas of solid continuum mechanics and earthquake engineering.

Tsuyoshi Ichimura received his B.E., M.E., and Ph.D. degrees from The University of Tokyo in 1998, 1999, and 2001. He became a professor at Earthquake Research Institute, The University of Tokyo, in 2019. He has been a director at the Research Center for Computational Earth Science since 2019. His research interests are in the areas of computational science and earthquake engineering.

Lalith Maddegedara received his B.Sc. degree from the University of Peradeniya, Sri Lanka, in 1999. He obtained his M.E. and Ph.D. degrees from The University of Tokyo, Japan, in 2000 and 2005, respectively. He became an associate professor at the Earthquake Research Institute, The University of Tokyo, in 2012. His research interests lie in computational mechanics, high-performance computing, and agent-based simulations for disaster applications.

Integrated Earthquake Simulation

M. Hori, T. Ichimura, and L. Maddegedara

CRC Press is an imprint of the
Taylor & Francis Group, an **informa** business

First edition published 2023
by CRC Press
6000 Broken Sound Parkway NW, Suite 300, Boca Raton, FL 33487-2742

and by CRC Press
4 Park Square, Milton Park, Abingdon, Oxon, OX14 4RN

CRC Press is an imprint of Taylor & Francis Group, LLC

© 2023 M. Hori, T. Ichimura, and L. Maddegedara

Reasonable efforts have been made to publish reliable data and information, but the author and publisher cannot assume responsibility for the validity of all materials or the consequences of their use. The authors and publishers have attempted to trace the copyright holders of all material reproduced in this publication and apologize to copyright holders if permission to publish in this form has not been obtained. If any copyright material has not been acknowledged please write and let us know so we may rectify in any future reprint.

Except as permitted under U.S. Copyright Law, no part of this book may be reprinted, reproduced, transmitted, or utilized in any form by any electronic, mechanical, or other means, now known or hereafter invented, including photocopying, microfilming, and recording, or in any information storage or retrieval system, without written permission from the publishers.

For permission to photocopy or use material electronically from this work, access www.copyright.com or contact the Copyright Clearance Center, Inc. (CCC), 222 Rosewood Drive, Danvers, MA 01923, 978-750-8400. For works that are not available on CCC please contact mpkbookspermissions@tandf.co.uk

Trademark notice: Product or corporate names may be trademarks or registered trademarks and are used only for identification and explanation without intent to infringe.

ISBN: 978-0-367-71185-6 (hbk)
ISBN: 978-0-367-71206-8 (pbk)
ISBN: 978-1-003-14979-8 (ebk)

DOI: 10.1201/9781003149798

Typeset in CMR10
by KnowledgeWorks Global Ltd.

Contents

List of Figures

List of Tables

Preface

This book provides the essential elements of the achievements that were made in a research project called *System for Integrated Simulation of Earthquake and Tsunami Hazard and Disaster*, which is a part of a national research program of *Research and Application Development in Post-K Project for Social and Scientific Important Tasks* during 2015 to 2019, organized by the Ministry of Education, Culture, Sports, Science and Technology, Japan. The research program was aimed at developing advanced numerical analysis methods which can take full advantage of Japanese supercomputers, K or Fugaku, and the project focused on developing a system which simulates all the processes of earthquake hazard and disaster by carrying out various numerical analysis methods, all of which make use of supercomputer.

From the viewpoint of academic research related to earthquakes, it is essential to develop a new method for estimating earthquake hazard and disaster in city and urban areas. While the present methods which are based on empirical relations are being used in practice, we need more reliable estimates for large earthquakes, which are rarely experienced. Also, we must consider the damages induced in digitized city and urban areas. The use of supercomputers enables us to develop a new method, which combines various numerical analyses of Earth Science and earthquake engineering, to realize the physics-based simulation of earthquake hazard and disaster. The development of such a system by combining various numerical analyses in the environment of supercomputers is challenging in the field of computer science and computational science. Therefore, it is no wonder that the above-mentioned research project was launched in the research program.

A unique characteristic of Earth science and earthquake engineering is that for the numerical simulation, we need an analysis model of high fidelity, in addition to an advanced numerical analysis method. The construction of a good analysis model for one structure requires a tedious task; hence, the construction of an analysis model for each structure located in an urban area will require huge tasks. However, various kinds of digital data are being accumulated, with which analysis models of certain quality can be automatically constructed. With advanced programs for such automated model construction, it is not impossible to construct analysis models for structures as well as the crust and the surface ground layers for the city and urban areas. The development of such programs is also challenging in the field of computational science.

Social simulation is needed to evaluate the indirect loss of the earthquake disaster, in addition to the direct loss that can be evaluated by the physics-based simulation. Although the number is limited, numerical analysis methods that use supercomputers are developed to simulate mass evacuation, traffic conditions, and post-disaster economic activities as an example of such social simulation. These analyses are still at the development stage but they could be a breakthrough for making a reliable estimate of earthquake hazard and disaster.

From the viewpoint of earthquake disaster mitigation, it is necessary to use advanced technologies for preparedness. Such an advanced technology is also useful to strengthen the resilience against earthquakes. A system of making a reliable estimate of earthquake hazard and disaster using supercomputers is a good candidate for such advanced technology. Using three elements, advanced numerical analyses that use the supercomputers, programs for automated model construction, and numerical analysis for social simulation, we present the examples of this new technology. The examples include the Tokyo Metropolis Earthquake and the Nankai Trough Earthquake, which are a clear and present threat to Japan.

In closing this preface, we would like to thank the following key members of the above-mentioned research projects, Prof. M. Asai (Kyusyu University), Dr. T. Hori (JAMSTEC), Dr. K. Iiyama (Kajima Corporation), Prof. J. Iryo (Tohoku University), Prof. Y. Sekimoto (The University of Tokyo), and Prof. M. Yokomatsu (Kyoto University), in the alphabetical order. The number of researchers involved in the project exceeds hundred, and we have selected researchers who represent research institutes involved in the project. However, we will never forget the efforts made by all the members.

Author Biography

Muneo Hori received his B.E. degree from The University of Tokyo, his M.S. degree from Northwestern University in 1985, and his Ph.D. degree from the University of California, San Diego, in 1987. He became a professor at the Earthquake Research Institute at The University of Tokyo, in 2001. He moved from the university to Japan Agency of Marine-Earth Science and Technology as a director general in 2019. His research interests are in the areas of solid continuum mechanics and earthquake engineering. He has been a managing editor of Journal of Earthquake Tsunami since 2011.

Tsuyoshi Ichimura received his B.E., M.E., and Ph.D. degrees from The University of Tokyo in 1998, 1999, and 2001. He became a professor at Earthquake Research Institute, The University of Tokyo, in 2019. He has been a director at the Research Center for Computational Earth Science since 2019. His research interests are in the areas of computational science and earthquake engineering.

Lalith Maddegedara received his B.Sc. degree from the University of Peradeniya, Sri Lanka, in 1999. He obtained his M.E. and Ph.D. degrees from The University of Tokyo, Japan, in 2000 and 2005, respectively. He became an associate professor at the Earthquake Research Institute, The University of Tokyo, in 2012. His research interests lie in computational mechanics, high-performance computing, and agent-based simulations for disaster applications.

1

Overview of Integrated Earthquake Simulation

CONTENTS

DOI: 10.1201/9781003149798-1

1.1 Background

Estimation of earthquake hazard and disaster is essential to plan efficient mitigation of damage to cities and urban areas caused by a probable earthquake. Here, earthquake hazard refers to the distribution of ground motion induced by an earthquake in the city and urban areas, and earthquake disaster refers to the total structural damage caused by ground motions; structural damage usually results in the failure of the city and urban area functions, such as transportation, energy supply, and communication, and thus generally included in earthquake disaster. Of course, the estimation of earthquake hazard and disaster for each region is not the goal of earthquake disaster mitigation but an element to achieve the same. However, if the estimates made are far from the actual earthquake hazard and disaster, we cannot mitigate earthquake disasters by any means. Thus, the estimate of regional earthquake hazard and disaster is an essential element for the earthquake disaster mitigation.

A conventional method to estimate regional earthquake hazard and disaster uses empirical relations, usually called an *attenuation equation* and a *fragility curve*. The attenuation equation is generally defined as a relation between the ground motion index (such as seismic index or maximum acceleration) and the distance from the epicenter of the earthquake. For a presumed earthquake, we can calculate the distribution of ground motion indices using the attenuation equation. The fragility curve is generally defined as a relation between the seismic index and the probability of structural damage. For a target structure, we can calculate the damage probability inputting the ground motion index (which is computed using the attenuation equation) into the fragility curve.

Various attenuation equations have been developed separately for many regions, to account for several factors such as the regional characteristics of earthquakes, the geological characteristics of the crust, and the geo-technical properties of surface ground. These factors influence the distribution of ground motions. When the use of seismographs was limited, it was highly important to collect ground motion data observed in an actual earthquake, and the attenuation equations were developed using the valuable data.

The fragility curves are developed for various structures, including wooden, steel, or reinforced concrete residential buildings, or buried pipelines used in lifelines of water, sewage, energy, and information. The curves usually depend on the construction year of the structure to account for the design codes and regulations according to which the structure was build. Roughly speaking, the ground motion index represents the external force, and the construction year corresponds to the structural seismic capacity.

It is true that the conventional method of using the attenuation equation and the fragility curve is a unique solution to estimate the earthquake hazard and disaster. Considering the number of structures located in a city and an urban area, it is a laborious task to compute the ground motion at its site and the possible damage for each structure, using the empirical relations. However, we often experienced that when a big earthquake hit the city and urban area, the resulting earthquake hazard and disaster did not agree with the estimate made by the conventional method, mainly because the actual earthquakes were different from the one presumed for the earthquake hazard and disaster estimate using the empirical relations. The disagreement was not negligible; a logarithm scale was used for the attenuation equation and the fragility curves were updated every time a large discrepancy was found between the estimated and the actual damages.

Considering the number of structures located in a city and an urban area, we must take advantage of the law of large numbers, i.e., whatever a probabilistic property is, the sum of probabilistic variables tends to obey a normal distribution.

According to the law, the sum of structural damages estimated by the fragility curve must provide an accurate value, even if the curve cannot predict the failure for one structure. This is the reason that the fragility curve is practically useful; as for each structure, the simple curve can only tell its damage in a probabilistic manner, but it can tell the accurate sum of the damage in a deterministic manner. The past reports of using the fragility curve did not support this usefulness of the fragility curve. It was suggested that the fragility curves used in the past were biased; i.e., the mean estimated by the curve was not accurate and the sum of the estimated damages could not match the sum of the actual damages. The methodology to determine a fragility curve using the actual data of ground motions and structural damages has been established. However, the methodology does not guarantee that the resulting curve is unbiased. There is no logical method to examine whether the fragility curve is biased or not mainly due to the lack of the quantity and quality of the actual data.

Recently, vast research has been made to develop physics-based simulation methods of the earthquake wave propagation processes. The quality of the numerical simulation in computing the processes is guaranteed, provided that a suitable analysis model of the crust and the surface ground is used and an *earthquake scenario* is given. The earthquake scenario refers to the fault mechanism, i.e., the rupture processes that take place on a fault plane. Such physics-based simulation can be applied to the earthquake hazard estimation, instead of the attenuation equation that is based on the observed data.

Higher spatial and temporal resolutions are needed to apply the physics-based simulation of the earthquake wave propagation processes to the earthquake hazard estimation, because the simulation methods developed are aimed at clarifying and understanding the earthquake wave propagation processes and computing in the length and temporal scales of 1,000 m and 1 s, respectively. Large-scale numerical computation is needed for the physics-based simulation, and modern computers of faster computing speed and larger compute memory are capable to perform such numerical computations in a relatively short time. The examples of such large-scale numerical computations are presented in Yoshimura *et al.* (2015). We need to develop a numerical simulation method which can take full advantage of such modern computers.

The use of physics-based simulation of the structural seismic response is an ordinary practice to examine the seismic safety of the new or existing important facilities, such as large-scale plants, transportation networks, ports, tunnels, and lifelines of water, sewage, energy, and information, as well as for large buildings including high rises. The structural seismic response analysis is physics-based, and the accuracy of simulating the seismic response is high; see, for instance, Fenves *et al.* (2011), Hori (2015), and Papadrakakis *et al.* (2017) for the use of numerical simulation in the field of earthquake engineering. Affordable commercial software is available for the structural seismic response simulation, making it a possible alternative of the fragility curve to estimate earthquake disaster. The reliability of the structural seismic response simulation is increased if ground motions synthesized by the earthquake wave propagation simulation are given to the target structure.

In applying the structural seismic response analysis to the earthquake disaster estimate, we must construct an analysis model for each structure located in a target city and urban area. If design charts of the structure are available, we can construct a reliable analysis model with which the reliability of the earthquake hazard estimate using the structural seismic response analysis becomes as high as that of the seismic design. It should be pointed out that we do not apply structural seismic response analysis for ordinary residential buildings. However, the structural seismic response analysis of such structures can make accurate estimate of structural damages even if a simple analysis model is used.

It is true that the earthquake disaster estimate using the structural seismic response analysis needs a large amount of numerical computation since the number of analyzed structures is of the order of 100,000 or 1,000,000. However, the numerical computation is much smaller than that of the earthquake hazard estimate using the physics-based simulation of the earthquake wave propagation processes. This is because simple analysis models are constructed for a majority of analysis models.

To construct an analysis model for structures located in a target city and urban area, we must use a large amount of digital data about the area. Even though the quality and quantity of the digital data are limited at this moment, we can well expect that more accurate and detailed data will be available in near future. It is essential to develop a technology which automatically constructs an analysis model using the digital data; correct information about structural configuration and material properties needs to be extracted from these data.

While digital data regarding the city and urban areas are used for the earthquake hazard estimate using the structural seismic response analysis,

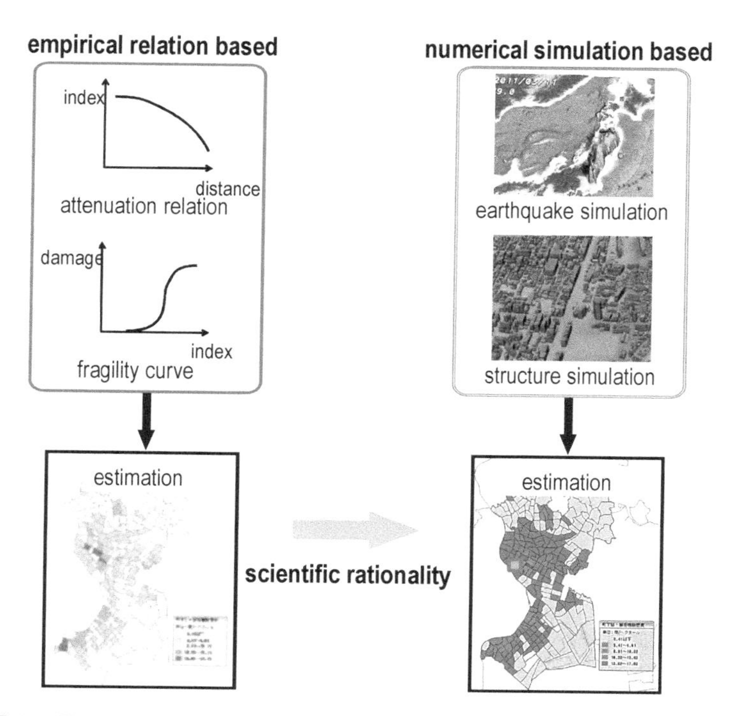

FIGURE 1.1: From empirical relations-based estimate to numerical simulation-based estimate.

vast geological and geo-technical data which have been accumulated over a few decades are used to construct an analysis model for the earthquake hazard estimate using the physics-based simulation. The analysis model includes an analysis model of the crust structure used for the simulation of earthquake wave propagation processes, or an analysis model of the surface ground layers used for the simulation of the earthquake wave amplification processes.

In closing this section, we summarize the discussions above as follows: we need a more reliable method of earthquake hazard and disaster estimate than the conventional empirical methods, and the physics-based simulation of earthquake wave propagation processes and the structural seismic response analysis are the candidates of such a method; see Fig. 1.1. Modern computers enable us to carry out large-scale numerical computation required for the regional simulation of high resolution for the earthquake hazard and disaster estimate. Digital data regarding the city and urban areas or geological and geo-technical data are available to construct analysis models for the regional simulation.

1.2 Scope

The objective of estimating the earthquake hazard and disaster is simple, as the estimation is used to ensure better preparedness before an earthquake. Better preparedness includes better design of the new structures and rational retrofitting of the existing structures, which must be cost-effective to mitigate structural damages. It also includes the organization of emergency response teams with stocks of resources required after an earthquake. The probability of an earthquake disaster occurrence is relatively low, and hence the cost-efficiency for preparedness is important for making a regional plan of earthquake disaster mitigation, which involves numerous structures, emergency teams, and stocked goods. Therefore, the earthquake hazard and disaster estimate must be sufficiently reliable to make cost-efficient preparedness. Cost-efficient preparedness contributes to the earthquake disaster mitigation, even when the magnitude of the actual earthquake disaster and hazard are different from the estimated ones.

It is natural to analyze all processes of earthquake hazard and disaster for reliable estimation, and the simulation-based method considered in the preceding section analyzes the processes of earthquake hazard using the physics-based simulation of earthquake wave propagation and the processes of earthquake disaster using the structural seismic response analysis. The conventional method using the empirical relations fails to fully analyze the earthquake hazard and disaster which are the sequence of small but numerous processes taking places in each part of a target city and urban area. Analyzing the processes of the earthquake hazard and disaster can result in a more reliable estimate if each process is analyzed with similar accuracy.

Integrated Earthquake Simulation (IES) realizes the estimation of earthquake hazard and disaster analyzing all the involved processes. Various pairs of an advanced numerical analysis method and analysis models are used in IES, and each pair performs the numerical analysis simulating a specific process of earthquake hazard and disaster in a city and an urban area; see[1] Hori (2018).

The overview of IES is presented in Fig. 1.2. As is shown, IES consists of the following three phases: 1) earth science simulation for the earthquake wave propagation processes; 2) earthquake engineering simulation of the structural seismic response processes; and 3) social simulation of the disaster reaction processes. The first phase mainly analyzes the crust, wherein an earthquake is generated on a fault plane, and earthquake waves propagate. Tsunami in the ocean is included in this phase for a given tsunami scenario. For a given

[1] See also the System for Integrated Simulation of Earthquake and Tsunami Hazard and Disaster 2020, which includes research achievements in IES development from 2015 to 2019. Although not implemented in IES, detailed explanation is given to the advanced numerical analyses of the particle simulation of Furuichi *et al.* (2017) or the economic simulation of Yokomatsu *et al.* (2015).

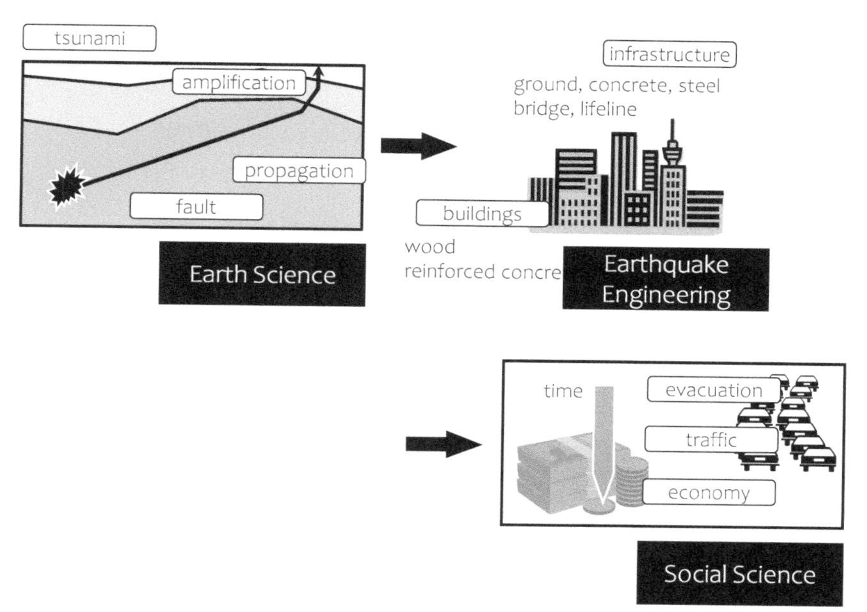

FIGURE 1.2: Overview of IES.

earthquake scenario, the first phase results in the estimation of earthquake hazard or a ground motion distribution. This output is used as an input to the second phase. The second phase analyzes a target city and urban area, wherein various structures are located. Indeed, IES is aimed at analyzing all structures using various methods of structural seismic response analysis. The output of the second phase is the estimation of earthquake disaster or a set of damaged structures. This output is used as an input to the third phase. In the third phase, a region or a country including a target city and urban area is analyzed using social simulations, such as post-disaster traffic flow simulation and economic simulations.

The inclusion of the third phase is a unique feature of IES. The earthquake disaster includes the direct loss of structural damages and indirect loss related to social and economic activities, which are induced by the direct loss as well as the failure of the city and urban area functionalities. Indirect loss often exceeds the direct loss in monetary terms. Hence, it is necessary to estimate the indirect loss; social simulation, studied in the field of social science, could be used for this purpose. The accuracy and reliability of such a social simulation are different from those of the physics-based simulation which includes the structural seismic response analysis. Linking the physics-based simulation to social simulation is challenging. We should keep in mind that the objective of estimating the indirect loss is the same as the objective of estimating the direct loss, i.e., better preparedness. The accuracy and reliability of social simulation must aim for this objective. For instance, the accuracy and reliability will be sufficient if we can select a better retrofitting plan among a few plans, by comparing the decrease in the direct and indirect losses which are computed according to each plan.

High accuracy is an essential element of numerical analysis. For more accurate numerical analysis performed by IES, it is necessary to develop better numerical methods and construct better analysis models. In addition to the accuracy of the numerical analysis, we must consider the following three points while analyzing the processes of earthquake hazard and disaster using IES: 1) the uncertainties of the earthquake hazard and disaster; 2) temporal and spatial resolutions of the analysis; and 3) the analysis resulting less errors while computing the total earthquake disaster.

The uncertainties of earthquake hazard and disaster are mainly related to an earthquake scenario. During a great earthquake, there will be severely shaken and damaged regions, which cannot be identified unless the correct earthquake scenario that matches the actual earthquake is used. The regions change with the earthquake scenario, but it is impossible to know the actual scenario of the earthquake. Thus, the following question arises in carrying out IES for the estimate of earthquake hazard and disaster:

> which is a good methodology to account for the uncertainties of earthquake hazard and disaster?

To address this question, we must consider maximum possible earthquake scenarios to avoid unconsidered situations in calculating the simulation-based estimate of earthquake hazard and disaster. Although probabilistic, this methodology provides the most scientifically reliable estimate of earthquake hazard and disaster. It was practically impossible to use this methodology for estimating earthquake hazard and disaster due to the limited computing power. Modern computers[2] enable us to make an estimate of earthquake hazard and disaster according to this methodology.

Since the objective of earthquake hazard and disaster estimate is better preparedness, the estimate is shared among various organizations. For instance, local governments need a rough estimate of damages to all structures in a region, while building owners need detailed estimation of damage to their assets. Thus, the estimate needs to be of high resolution, additionally, covering the entire region. It is imperative to sum up results of high-resolution estimates for the reliable estimate of regional earthquake damage. This corresponds to the actual earthquake disaster, which is the sum of damage of an individual structure induced by the ground motion at the site. To achieve high spatial resolution in the second phase of IES, we must have high temporal resolution as well, because the spatial and temporal resolutions are closely related in the first phase of IES. An earthquake wave component of longer

[2]The digital data of many city and urban area elements, including dynamic data of the social and economic activities in addition to static data of the residential buildings and infrastructures, such as lifelines, road networks, railway networks, and port facilities, are available presently. They are the sources for constructing reliable analysis models for IES.

period (or low frequency) has[3] a longer wave length. To have a structure-wise spatial resolution in the second phase, we must have a temporal resolution shorter than 1 s (close to 0.1 s) in the first phase.

In summing up high-resolution estimates of IES, we must pay attention to the law of large numbers, as explained in the preceding section. If the high-resolution estimate is not biased, i.e., the estimate of the mean is accurate, the law of large numbers guarantees that despite an inaccurate estimate for each structure, the sum of individual estimates tends to be more accurate, suggesting an unbiased estimate. A physics-based simulation analyzing all the processes of earthquake hazard and disaster can exemplify this point if a good numerical analysis method and a good numerical analysis model are used. The physics-based simulation that is implemented in IES is a logical choice for an unbiased estimation. It is true that such a physics-based simulation needs a set of parameters and that certain errors are included in determining them from limited data. Thus, we must improve the determination of the model parameters to reduce the bias. A way of improving the parameter determination is to collect all the data related to the structure, including digital materials of design charts and results of structural health monitoring. As for social simulations, it is much more difficult to make an unbiased estimation; they must not be treated like a physics-based simulation in terms of accuracy, resolution, and bias. At this moment, the social simulations which are implemented in IES are aimed at providing the information about selecting a better retrofitting plan among a few plans.

1.3 Key features

In this monograph, integrated simulation is defined as the sequential execution of a set of numerical analyses, in which the output of a numerical analysis is used as the input of the next numerical analysis. Combining the analyses this way is straightforward without requiring any special process except for converting the output results of the preceding numerical analysis into the input data of the next one. IES is an integrated simulation, analyzing all the processes of earthquake hazard and disaster in a city and an urban area with various numerical analyses applied to each of these processes.

IES carries out an integrated simulation, including regional simulation for earthquake hazard estimate and structural simulation for earthquake disaster estimate. The simulation contains the following two key features: 1) the utilization of *high-performance computing* (HPC) for numerical analysis methods implemented in IES; and 2) the automated model construction applied to the

[3]If the wave velocity is v, an earthquake wave component of period t has a wave length of vt.

crust, ground surface, structures, and social and economic activities. Here, HPC refers to parallel computing or computing accelerated by GPUs or vector processors. The automated model construction is used to make an analysis model of high fidelity for the first phase and numerous analysis models for the second and third phases; a numerical analysis model must be constructed for each structure or road segment in a city and an urban area.

Since IES is aimed at estimating the earthquake hazard and disaster, each numerical analysis must be applied to a regional scale problem with optimally fine spatial and temporal resolution. For instance, the size, spatial resolution, and time increment of a surface ground analysis model used in the earthquake wave amplification analysis are of the order of 10,000 m, 1 m, and 0.01 s, respectively, to cover an earthquake lasting for 100 s. The numerical analysis of this model requires intensive computing, and HPC is primarily important for the numerical analysis method to complete with the computation in a short time. HPC is needed for the structural seismic response analysis, because a regional problem of analyzing all structures located in a city and an urban area must be solved. The number of the structures analyzed is at least of the order of 100,000. Even if a relatively small analysis model is used for a structure, the total computing time becomes unacceptable unless HPC is used.

Automated model construction is needed for the crust, which consists of geological layers of distinct configuration and material properties. An analysis model of high fidelity must be used to accurately analyze the wave propagation processes in the first phase of IES. As will be explained later, such a high-fidelity analysis model cannot be constructed manually. Automated model construction is more important for the second and third phases of IES, wherein structural seismic responses and post-disaster social responses are analyzed. As mentioned, the number of the target structures exceeds 100,000 and manual construction of such a large number of analysis models is unrealistic.

We briefly review the integrated simulation developed in the field of earthquake engineering. The pioneering work was reported by Hori 2018 as the concept of integrated simulation was introduced to estimate the earthquake hazard and disaster using physics-based simulation. The target of this work was a small part of a city in which dozens of residential houses were located. The architecture of the integration is changed from the mediator which links components in the system to the layer structure which connects components located in different layers. Lu and Guan (2017) developed an integrated simulation that used actual data, focusing on structural seismic response analysis in China. The use of the actual data for structural seismic response analysis contributed to higher reliability of the simulation results. Sahin *et al.* (2016) developed an integrated system for Istanbul, Turkey, adding components of engineering seismology. The developed system covers several parts of Istanbul using available data. In USA, so-called regional simulation is a key issue as an application of HPC; see Fenves *et al.* (2011). Advanced programs that are enhanced with the HPC capability have been developed for estimating earthquake hazard and disaster in a wide region with high spatial resolution; see Yoshimura *et al.* (2015).

1.3.1 Utilization of HPC

The physics-based simulation of earthquake hazard and disaster, which correspond to the first and second phases of IES, is recognized as solving a wave equation of solid continuum mechanics for the crust, surface ground, or structures. Since the wave equation is a set of second-order partial differential equations for the three components of a displacement function, we must use HPC to numerically solve this equation in a sufficiently large domain with sufficiently fine spatial and temporal resolutions; the amount of numerical computation increases as the domain size becomes larger or the resolution become finer. The use of HPC is important for completing the numerical computation in a designated time, provided that a suitable computing environment is used.

A set of numerical analysis programs, each of which is developed for its own specific purpose, are implemented in the current system of IES. All the numerical analysis programs are enhanced with the HPC capability when they are implemented in IES. The tuning of each program is needed to increase its performance of computing as well as to be scalable so that computing time can be estimated for a given computer. It should be pointed out that compared with implementing various numerical analysis programs into IES, implementing one general-purpose finite element method (FEM) into IES is much easier as it is applicable to solve the wave equation for various analysis models for the crust, surface ground, or many kinds of structures; thus, such a general-purpose FEM tends to be more frequently used in earthquake engineering. It is straightforward to enhance FEM with the HPC capability because most of the computing time of FEM is spent on solving a matrix equation that is derived from the wave equation, and only this part should be enhanced with the HPC capability. We explain this enhancement in the next chapter.

The use of HPC is inevitable to analyze a large-scale analysis model which requires immense numerical computation. Here, the scale of the analysis model refers to the degree-of-freedom which is regarded as unknowns to be solved in the numerical computation. As the scale of the analysis model becomes higher, the fidelity of the analysis model, which is evaluated in terms of the detailedness of the configuration and the material properties, becomes higher, and a higher fidelity analysis model generates more reliable predictions in the physics-based simulation[4] in the sense that it yields a more accurate numerical solution of the wave equation.

The construction of a high-fidelity analysis model requires detailed information about a target body. However, the quality and quantity of available data are limited for underground and structures. This results in the

[4]Fluid-solid interaction problems such as tsunami or liquefaction are also included in the physics-based simulation. These problems couple the wave equation of the solid to other equations of the fluid; see, for instance, Asai *et al.* (2016) and Chen *et al.* (2019).

uncertainty of prediction, even if a large-scale model is analyzed by using HPC. *Ensemble computing*[5] is a method to quantify the uncertainties in predicting the ground motion and structural seismic response. This computing uses a set of analysis models which are constructed for one target body by stochastically changing uncertain parameters of the analysis model. While analyzing all the analysis models for one body needs a large amount of numerical computation, it provides the most reliable stochastic estimate.

1.3.2 Automated model construction

For the first phase of IES, we must construct analysis models for the crust. Various observation data are available in the field of seismology and geodesy. The observation data which are obtained using a network of seismographs are generally used to construct a crust model. We must convert the crust model to an analysis model which is used as an input of a numerical analysis method of the earthquake wave propagation. This conversion must be automatically made, since the numerical analysis model becomes large scale as its degree-of-freedom is of the order of 1,000,000 or more.

In converting the curst model into an analysis model automatically, we must pay attention to the fidelity of the analysis model. An analysis model of high or low fidelity can be constructed from the crust model, changing the scale of the analysis model. Since the intricate configurations of the crust model influence the complex processes of earthquake wave propagation, it is essential to analyze an analysis model of high fidelity, which inevitably becomes of large scale. The automated construction of a high-fidelity analysis model (or large-scale analysis model) is important for the wave propagation analysis of IES.

The digital data of city areas are being rapidly accumulated, as they are regarded as fundamental information about the areas. The quantity as well as the quality of such digital data increase, and it is possible to construct analysis models for structures using the digital data for the second and third phases of IES. At the present, three-dimensional models[6] have been constructed for all structures located in several big cities in the world using the digital data. Automated model construction can be developed so that digital data of city areas (if available, digital data of urban areas) are converted to a suitable analysis model of a structure.

[5]Ensemble computing or capacity computing for solving numerous models of similar kinds is regarded as another category of HPC application, besides the capability computing of solving a large-scale analysis model.

[6]These models are called a *digital twin* of a city area. While analysis models used in IES are for numerical computation, a digital twin is a model mainly used for visualization and usually used as a three-dimensional map. In general, a model for visualization needs data about external configuration, but a model for numerical analysis needs additional data, such as data about internal configurations and structural members and data about material properties.

For automated model construction, it is necessary to convert various digital data into an analysis model used for a specific numerical analysis, by inter- or extrapolating the missing information about the configuration and material properties of the structures. As for a residential building, for instance, data conversion must include a choice of a suitable analysis model, which ranges from a simple mass-spring model to a complicated solid element model, evaluating the quality and quantity of the available digital data.

It should be pointed out that inter- or extrapolating the missing information has a limitation, and some uncertainty remains in automatically constructing an analysis model. Even if a suitable analysis model is chosen, it is impossible to construct a perfect analysis model unless design charts are used. Automated model construction must account for the uncertainty of the analysis model, or the uncertainty of model parameters which cannot be accurately determined from the available digital data. The simplest solution for the uncertainty is ensemble computing that uses a set of stochastically generated analysis models. A probabilistic distribution[7] of the model parameters needs to be presumed.

In the second and third phases of IES, two numerical analyses are applied to structures. For example, the structural seismic response analysis is applied to all structures in the second phases, and the spatial distribution of debris of the damaged structures are analyzed in the third phase to carry out the traffic flow simulation that uses a network of damaged road segments. As is seen, digital data of one structure are used to construct two analysis models. While it must combine different digital data for one structure, automated model construction must convert combined digital data to different analysis models for different numerical analyses.

[7] A normal distribution is typically used, and no correlation is assumed for two or many model parameters.

2

Applications Implemented in Integrated Simulation

CONTENTS

DOI: 10.1201/9781003149798-2

2.1 Finite element method of solving wave equation

This section explains the analysis methods of wave propagation during earthquakes from the faults to the ground surface. At present, the process of wave propagation is often divided into seismic wave propagation, which deals with wave propagation from the fault to the engineering basin, and ground motion amplification, which is caused by a ground structure shallower than the engineering base. Seismic wave propagation and ground motion amplification are considered as wave propagation in solids; thus, numerical simulations of wave propagation are effective for their analysis, and many studies have been conducted (e.g., Poursartip *et al.* 2020). Here, we describe the numerical simulation of seismic wave propagation and ground motion amplification using the finite element method (Zienkiewicz and Taylor 2005), which can deal with the geometry of the model accurately and satisfy the stress-free boundary condition on the surface analytically because the complex geometry of the model strongly affects the numerical results. Since in the case of explicit time integration, the Courant condition becomes severe due to the very small finite elements that are generated locally when modeling complex geometries, implicit time integration is used. In addition, we will use low-order unstructured tetrahedral finite elements, which are typically used in mechanical systems and are excellent for modeling complex geometries.

2.1.1 Governing equation

The governing equation for wave propagation in solids

$$\frac{\partial}{\partial x_i}\left(c_{ijkl}(\mathbf{x})\frac{\partial}{\partial x_l}u_k(\mathbf{x},t)\right) - \rho(\mathbf{x})\ddot{u}_j(\mathbf{x},t) = f_j(\mathbf{x},t), \quad (2.1)$$

is considered. For simplicity, we assume a Cartesian coordinate system consisting of x_1, x_2, and x_3. Here, $\mathbf{x}$ and t are the spatial position and time, and $\frac{\partial}{\partial x_i}$ and $\ddot{}$ represent the partial derivative in the direction of x_i and the (second-order) time derivative, respectively. Also, c_{ijkl}, u_i, f_i, and ρ represent a heterogeneous elasticity tensor, displacement in the i-direction, external force in the i-direction, and density, respectively. Seismic wave propagation is often modeled as a linear wave field, where c_{ijkl} in Eq. (2.1) does not vary with time t. Because there are still many unknowns in the fault processes in the high-frequency region and because the crustal structure model is not yet sufficiently obtained in high resolution, numerical simulations are performed for the low-frequency component (about 1 Hz or less). The high-frequency component (above 1 Hz) is often estimated by the stochastic Green's function method (Boore 1983) or the empirical Green's function method (Irikura 1986), and then combined with a matching filter to obtain a broadband waveform. Ground motion amplification is often modeled as a nonlinear wavefield where c_{ijkl} in Eq. (2.1) varies with the constitutive law of the soil (i.e., nonlinear relationship between strain and stiffness) for a broadband input waveform set on an engineering basin. Note that the actual problem is solved by adding a damping term to Eq. (2.1) because there are various damping factors.

2.1.2 Boundary condition

Since the finite element method is an analysis method for closed domains, when numerical simulations of seismic wave propagation are performed, it is necessary to set the boundary conditions so that the waves generated inside the domain can properly penetrate the boundary. A typical boundary condition is the absorbing boundary condition (Clayton and Engquist 1977) that prevents reflected waves by decomposing the governing equations into different directions of wave propagation. Specifically, the one-dimensional wave equation for a scalar value of u,

$$E\frac{\partial^2 u}{\partial x^2} = \rho\frac{\partial^2 u}{\partial t^2} \qquad (0 \leq x \leq 1), \tag{2.2}$$

is considered as an example. Here, E and ρ are Young's modulus and density, respectively. The absorbing boundary condition is set at $x = 1$ for Eq. (2.2) to transmit outward at $x = 1$ when the wave is input at $x = 0$. With the wave speed $c = \sqrt{E/\rho}$, Eq. (2.2) is decomposed to

$$\frac{\partial u}{\partial x} + \frac{1}{c}\frac{\partial u}{\partial t} = 0, \tag{2.3}$$

$$\frac{\partial u}{\partial x} - \frac{1}{c}\frac{\partial u}{\partial t} = 0. \tag{2.4}$$

Since Eq. (2.3) has only waves traveling in the positive direction of x, and Eq. (2.4) has only waves traveling in the negative direction of x, if Eq. (2.3) is applied at $x = 1$ as the boundary condition, it is expected to eliminate the waves reflected from the boundary edge at $x = 1$. This boundary condition allows for the complete elimination of waves from the boundary in the one-dimensional wave equation Eq. (2.2). On the other hand, for a higher dimensional wave equation, it is difficult to completely eliminate the waves from the boundary because of the difficulty in complete decomposition such as Eq. (2.2) to Eqs. (2.3) and (2.4). Therefore, additional methods are used for eliminating the reflected waves from the boundary (e.g., selectively applying attenuation to the calculation results in a certain area or an absorption band from the boundary; see Cerjan *et al.* (1985). In addition, the perfect matched layer method (Berenger 1994), which is a little more complicated to implement but has even better performance, is also used.

In ground motion amplification, it is necessary to introduce the effect that the ground is continuous in addition to the condition that the waves penetrate at the boundary. There are several methods; it is handled as follows in one of the typical methods. First, a one-dimensional structure similar to that of the side surface of the 3D finite element model is created; then a one-dimensional amplification analysis is performed for the input seismic motion to obtain the displacement response $u^0(\mathbf{x}, t)$, and the force is applied to the boundary so that this response is reproduced in 3D analysis. If the ground structure is horizontally stratified, the response at the ground surface will be uniform because the continuity condition of the ground is completely satisfied. If this is not the case, the disturbance waves caused by the inhomogeneity of the target will be reflected at the boundary, and it is necessary to allow them to pass through. To achieve this, an absorbing boundary is applied to the time derivative of the difference between the response $u(\mathbf{x}, t)$ and $u^0(\mathbf{x}, t)$ from the three-dimensional analysis at the side and bottom surfaces.

2.1.3 Solution algorithm

The solution algorithm for seismic wave propagation and ground motion amplification is obtained by discretizing the time domain using Newmark's β method ($\delta = 1/2$ and $\beta = 1/4$), which is an implicit time integral commonly used in earthquake engineering, and the spatial domain using the finite element method based on the Galerkin method. Both methods will have to solve large-dimensional matrix equations but they have the great advantage of preventing instability in the analysis. Since Newmark's β method is a kind of Taylor expansion, even though the analysis is stable, the time step increment must be set to be small enough for the solution to converge. In order to obtain a convergent solution, the spatial discretization should be at least 8–10 elements per wavelength at the frequency of interest when using empirical first-order elements, or 4–5 elements per wavelength at the frequency of interest when using second-order elements. Note that since the surface waves have a wave speed slower than the body wave, more elements per wavelength are needed. For checking the accuracy of the numerical results, a semi-analytical solution for point sources on horizontally stratified structures (Hisada 1994) is used for seismic wave propagation, while a semi-analytical solution for plane wave incidence on horizontally stratified structures is used for ground motion amplification. After checking the accuracy with a semi-analytical solution, a more complex model is used to check whether a convergent solution can be obtained by gradually using elements of smaller sizes.

The discretized governing equation of seismic wave propagation is

$$\left(\frac{1}{dt^2}\mathbf{M} + \frac{1}{dt}\mathbf{C} + \mathbf{K}\right)\mathbf{u}^n = \mathbf{F}^n + \mathbf{C}\mathbf{v}^{n-1} + \mathbf{M}\left(\mathbf{a}^{n-1} + \frac{4}{dt}\mathbf{v}^{n-1}\right), \tag{2.5}$$

where

$$\begin{cases} \mathbf{v}^n = -\mathbf{v}^{n-1} + \frac{2}{dt}(\mathbf{u}^n - \mathbf{u}^{n-1}), \\ \mathbf{a}^n = -\mathbf{a}^{n-1} - \frac{4}{dt}\mathbf{v}^{n-1} + \frac{4}{dt^2}(\mathbf{u}^n - \mathbf{u}^{n-1}), \end{cases} \tag{2.6}$$

where $\mathbf{u}$, $\mathbf{v}$, $\mathbf{a}$, and $\mathbf{F}$ are the displacement, velocity, acceleration, and external force vectors, respectively. $\mathbf{M}$, $\mathbf{C}$, and $\mathbf{K}$ are the mass, damping, and stiffness matrices, respectively. dt and n represent the number of time increments or time steps. The Rayleigh damping matrix is used as $\mathbf{C}$, and the element damping matrix $\mathbf{C}_e$ is used with the element mass matrix $\mathbf{M}_e$ and the element stiffness matrix $\mathbf{K}_e$ given below.

$$\mathbf{C}_e = \alpha \mathbf{M}_e + \beta \mathbf{K}_e.$$

α and β are determined by solving the following least-squares problem.

$$minimize \left[\int_{f_{min}}^{f_{max}} \left(h - \frac{1}{2} \left(\frac{\alpha}{2\pi f} + 2\pi f \beta \right) \right)^2 df \right].$$

Here, f_{max}, f_{min}, and h are the maximum and minimum frequencies, and the damping constants to be analyzed, respectively. First, the external force term due to the earthquake source is set to $\mathbf{F}^n$, and the matrix Eq. (2.5) is solved using the values up to the $n-1$ time step to obtain the displacement $\mathbf{u}^n$ at the n time step. This is used to obtain $\mathbf{v}^n$ and $\mathbf{a}^n$ by Eq. (2.6). The above is repeated step-by-step to obtain the time history response. Note that $\mathbf{C}$ includes not only internal damping but also the damping of the absorbing boundaries at the bottom and the sides of the domain.

Assuming that the time increment size is sufficiently small, the discretized governing equation of the ground motion amplification is

$$\left(\frac{4}{dt^2} \mathbf{M} + \frac{2}{dt} \mathbf{C}^n + \mathbf{K}^n \right) \delta \mathbf{u}^n = \mathbf{F}^n - \mathbf{Q}^{n-1} + \mathbf{C}^n \mathbf{v}^{n-1} + \mathbf{M} \left(\mathbf{a}^{n-1} + \frac{4}{dt} \mathbf{v}^{n-1} \right), \tag{2.7}$$

where

$$\begin{cases} \mathbf{Q}^n = \mathbf{Q}^{n-1} + \mathbf{K}^n \delta \mathbf{u}^n, \\ \mathbf{u}^n = \mathbf{u}^{n-1} + \delta \mathbf{u}^n, \\ \mathbf{v}^n = -\mathbf{v}^{n-1} + \frac{2}{dt} \delta \mathbf{u}^n, \\ \mathbf{a}^n = -\mathbf{a}^{n-1} - \frac{4}{dt} \mathbf{v}^{n-1} + \frac{4}{dt^2} \delta \mathbf{u}^n, \end{cases} \tag{2.8}$$

where $\delta \mathbf{u}$, $\mathbf{u}$, $\mathbf{v}$, $\mathbf{a}$, and $\mathbf{F}$ are incremental displacement, displacement, velocity, acceleration, and external acceleration, respectively, and $\delta \mathbf{u}$, $\mathbf{u}$, $\mathbf{v}$, $\mathbf{a}$, and $\mathbf{F}$ are the incremental displacement, displacement, velocity, acceleration, and external force vectors, respectively. $\mathbf{M}$, $\mathbf{C}$, and $\mathbf{K}$ are the mass, damping, and stiffness matrices, respectively. dt and n represent the number of time increments or time steps. Since the stiffness and damping vary with the time step, the Rayleigh damping matrix is used as $\mathbf{C}$, and the element damping matrix $\mathbf{C}_e^n$ is calculated using the element mass matrix $\mathbf{M}_e$ and the element stiffness matrix $\mathbf{K}_e^n$.

$$\mathbf{C}_e^n = \alpha \mathbf{M}_e + \beta \mathbf{K}_e^n.$$

α and β are determined by solving the following least-squares problem.

$$minimize \left[\int_{f_{min}}^{f_{max}} \left(h^n - \frac{1}{2} \left(\frac{\alpha}{2\pi f} + 2\pi f \beta \right) \right)^2 \mathrm{d}f \right].$$

Here, f_{max}, f_{min}, and h^n are the damping constants for the maximum frequency, minimum frequency, and the time step n to be analyzed. The specific calculation procedure considering the stiffness and the damping changes due to the nonlinear constitutive law of the ground is to repeat the following.

1. Using the strains calculated in the $n-1$-th time step, the stiffness and damping at the n-th time step are evaluate using the constitutive law of the ground.
2. Using the stiffness and damping of the n-th time step, $\mathbf{K}^n$ and $\mathbf{C}^n$ are determined.
3. The equation in Eq. (2.7) is solved to obtain $\delta\mathbf{u}^n$ and each value is updated in Eq. (2.8).

Note that $\mathbf{C}^n$ includes the absorbing boundary conditions applied to the sides and the bottom, and $\mathbf{F}^n$ includes the input wave from the bottom and the force from the ground continuity condition.

2.1.4 Solution algorithm with high-performance computing

Equations (2.5) and (2.7) are both matrix equations with a large number of degrees of freedom, and their solution requires ingenuity with high-performance computing (HPC). Since the target matrices are sparse, an iterative solver is used. Another reason for using the iterative solver is that the convergence is relatively good since the main components are concentrated near the diagonal of the target matrices. Many methods have been proposed as the iterative solver with Krylov subspaces, and the conjugate gradient method (Saad 2003) for positive definite symmetric matrices is one of the most representative ones. Since the matrix equation obtained by the finite element method is often a positive definite symmetric matrix, the conjugate gradient method is often used to calculate the response of large-scale finite element models. The key mechanism of the conjugate gradient method is shown below. Consider the solution of $\mathbf{Ax} = \mathbf{f}$ as the following equivalent minimization problem for $\mathbf{x}$.

$$\Phi = \frac{1}{2}(\mathbf{x}, \mathbf{Ax}) - (\mathbf{f}, \mathbf{x}). \tag{2.9}$$

In the steepest descent method, which is a robust minimum value search method, the solution is updated sequentially in the direction that reduces the error the most, starting from an appropriate initial value $\mathbf{x}_0$, as follows:

$$\mathbf{x}_{k+1} = \mathbf{x}_k + \gamma\mathbf{d}_k,$$

where $\mathbf{d}_k = -\nabla\Phi(\mathbf{x}_k)$. In the case of Eq. (2.9), $\mathbf{d}_k = \mathbf{f} - \mathbf{Ax}_k$. While changing γ $(0 < \gamma << 1)$ step by step, efficient minimization is attempted but there is a weakness in that the number of iterations to reach the minimum value cannot be suppressed. On the other hand, for the conjugate vector $\mathbf{p}_i$ for $\mathbf{A}$, we have

$$(\mathbf{p}_i, \mathbf{Ap}_j) = 0 \quad (i \neq j).$$

Algorithm 1 Conjugate gradient method to find the numerical solution of the matrix equation $\mathbf{Ax} = \mathbf{f}$ with relative error $(||\mathbf{r}_i||/||\mathbf{f}|| < \epsilon)$.

$\mathbf{r}_0 \Leftarrow \mathbf{f} - \mathbf{A}\mathbf{x}_0$
$\mathbf{p}_0 \Leftarrow \mathbf{r}0$
$i \Leftarrow 0$
while $(\|\mathbf{r}_i\|/\|\mathbf{f}\| \geq \epsilon$ **do**
  $\alpha_i \Leftarrow \frac{(\mathbf{r}_i,\mathbf{p}_i)}{(\mathbf{p}_i,\mathbf{A}\mathbf{p}_i)}$
  $\mathbf{x}_{i+1} \Leftarrow \mathbf{x}_i + \alpha_i \mathbf{p}_i$
  $\mathbf{r}_{i+1} \Leftarrow \mathbf{r}_i - \alpha_i \mathbf{A}\mathbf{p}_i$
  $\beta_i \Leftarrow \frac{(\mathbf{r}_{i+1},\mathbf{A}\mathbf{p}_i)}{(\mathbf{p}_i,\mathbf{A}\mathbf{p}_i)}$
  $\mathbf{p}_{i+1} \Leftarrow \mathbf{r}_{i+1} - \beta_i \mathbf{p}_i$
  $i \Leftarrow i + 1$
end while

Using the orthogonality property that any solution can be expanded as

$$\mathbf{x} = \sum_{i=1}^{n} \alpha_i \mathbf{p}_i,$$

where $\alpha_i = \frac{(\mathbf{p}_i,\mathbf{f})}{(\mathbf{p}_i,\mathbf{A}\mathbf{p}_i)}$, this shows the strong characteristic that if there is an algorithm to find $\mathbf{p}_i$ sequentially, it will reach the correct answer in at most n iterations; the conjugate gradient method is just such an algorithm. Specifically, we set the initial solution as $\mathbf{x}_0$, the initial residual as $\mathbf{r}_0 = \mathbf{f} - \mathbf{A}\mathbf{x}_0$, and the initial conjugate vector as $\mathbf{p}_0 = \mathbf{r}_0$.

$$\begin{aligned}
\mathbf{x}_1 &= \mathbf{x}_0 + \alpha_0 \mathbf{p}_0, \\
\mathbf{r}_1 &= \mathbf{f} - \mathbf{A}\mathbf{x}_1 = \mathbf{r}_0 - \alpha_0 \mathbf{A}\mathbf{p}_0, \\
\mathbf{p}_1 &= \mathbf{r}_1 - \beta_0 \mathbf{p}_0,
\end{aligned}$$

where $\alpha_0 = \frac{(\mathbf{r}_0,\mathbf{p}_0)}{(\mathbf{p}_0,\mathbf{A}\mathbf{p}_0)}$ and $\beta_0 = \frac{(\mathbf{r}_1,\mathbf{A}\mathbf{p}_0)}{(\mathbf{p}_0,\mathbf{A}\mathbf{p}_0)}$. By repeating the above procedure, we obtain the conjugate vector. Since we try to obtain a conjugate vector $\mathbf{p}_*$ that is close to the steepest direction $\mathbf{r}_*$, we can see that the algorithm is close to the steepest descent method and can find the solution in as few as n iterations. In practice, Algorithm 1 finds the solution by updating the initial solution $\mathbf{x}_0$ until the relative error $(\|\mathbf{r}_i\|/\|\mathbf{f}\|)$ is smaller than the positive number $\epsilon \ll 1$.

The number of iterations of the conjugate gradient method depends on the maximum/minimum eigenvalues of $\mathbf{A}$ and the distribution of the eigenvalues, and the number of iterations decreases if $\mathbf{A}$ is closer to the unit matrix. Therefore, in order to reduce the number of iterations, $\mathbf{A}$ is multiplied by a matrix $\mathbf{M}$, for example, $\mathbf{MA}$. Such a process is called preconditioning, and $\mathbf{M}$ is called the preconditioning matrix. Typical preconditioning methods are point Jacobi, which extracts the diagonal terms of $\mathbf{A}$, block Jacobi, which extracts the blocks of diagonal terms of $\mathbf{A}$, and incomplete LU factorization,

Algorithm 2 Calculation of $\mathbf{f} = \mathbf{K}\mathbf{u}$ using the element-by-element method, where cny_i^{ie} is the global node number of the i-node of element ie, $\mathbf{Ke}^{ie}$ is the element stiffness matrix of element ie, and $\mathbf{ut}$ and $\mathbf{ft}$ are 30-dimensional temporary vectors.

```
for ie = 1, ··· , ne do
  for i = 1, ··· , 10 do
    for ii = 1, ··· , 3 do
      ut(3(i − 1) + ii) ⇐ u(3(cny_i^ie − 1) + ii)
    end for
  end for
  ft = Ke^ie ut
  for i = 1, ··· , 10 do
    for ii = 1, ··· , 3 do
      f(3(cny_i^ie − 1) + ii) = f(3(cny_i^ie − 1) + ii) + ft(3(i − 1) + ii)
    end for
  end for
end for
```

which performs LU decomposition of $\mathbf{A}$ without allowing fill-in. It is desirable to select a preconditioning method that takes into account the trade-off between the effect of preconditioning on reducing the number of iterations and the cost in the construction of the preconditioning matrix.

When performing finite element analysis using a solution method such as the conjugate gradient method, it is sometimes difficult to keep $\mathbf{A}$ in memory for large degrees of freedom problems, even though it is a sparse matrix. In such cases, the element-by-element method (Winget and Hughes 1985) can reduce the required memory amount, which is a method to perform equivalent analysis without keeping $\mathbf{A}$ in the memory. For simplicity, the calculation method of $\mathbf{f} = \mathbf{K}\mathbf{u}$ for static analysis without boundary conditions is shown in Algorithm 2. Here, we consider a finite element model consisting of ne tetrahedral quadratic elements, where the number of nodes is n, $\mathbf{u}$ has $3n$ dimensions, $u(3(i-1)+ii)$ is the displacement component of i nodes in the ii direction, and $\mathbf{K}$ is the overall stiffness matrix.

By extending the fundamental solution method described above with further HPC methods (e.g., mixed-precision arithmetic, utilizing sparsity of the solution space while taking into account the characteristics of the architecture and the system of computers), fast computation of large-scale problems using massive parallel computation becomes possible (e.g., Ichimura *et al.* 2014 and Ichimura *et al.* 2015). Furthermore, studies to realize faster solvers considering more data science-oriented approaches have been gradually reported. For example, there is a study that realized the speed-up of the entire solver by estimating the local ill-posedness of the target system using neural networks (e.g., Ichimura *et al.* 2018), and a study that realized the speed-up of the solver by improving the convergence of the solver with an advanced preconditioner using neural networks (e.g., Ichimura *et al.* 2020).

2.2 Structural seismic response analysis

This section presents the structural seismic response analysis implemented in the present system of IES. Since it forms a core element in earthquake engineering simulation, we start from the theoretical foundation of the analysis and explain the analysis that uses a multi-degree-of-freedom model for a residential building.

2.2.1 Foundation of structural seismic response analysis

In this monograph, we consider the foundation of structural seismic response analysis as continuum mechanics, which is based on Newtonian mechanics. Structural mechanics is generally considered as the foundation, but we regard structural mechanics as a *smart* mathematical approximation of continuum mechanics; see Appendix A. Continuum mechanics derives a set of coupled four-dimensional partial differential equations for a displacement vector function, which cannot be solved without modern computers. Structural mechanics is used to approximately solve this set, by converting the partial differential equations into a much simpler form so that a suitable approximate solution of continuum mechanics can be obtained by manual calculation or small computers; see Hori *et al.* (2014).

Approximation made in structural mechanics is mathematical, and it is the restriction[1] of a function that is used to express displacement or other physical fields. As shown in Fig. 2.1, a function space of structural mechanics is a subset of continuum mechanics because all functions used in structural mechanics belong to the function space of continuum mechanics. However, there are functions in continuum mechanics that are not treated in structural mechanics. We consider that the function space of the mass-spring system is a subset of the function space of structural mechanics, and that the mass-spring system is another mathematical approximation of continuum mechanics; much greater restriction is posed for a function of the displacement for the mass-spring system. In Fig. 2.1, a solution within each of the three function spaces is plotted. The solution in the function space of continuum mechanics is exact, and the other two solutions in the function space of structural mechanics and

[1] Hori *et al.* (2014) propose a meta-modeling theory that derives consistent mathematical problems of structural mechanics or the mass-spring system from a common Lagrangian of continuum mechanics. This theory is regarded as a simple application of the variational principle but provides a mathematical interpretation of structural mechanics based on continuum mechanics.

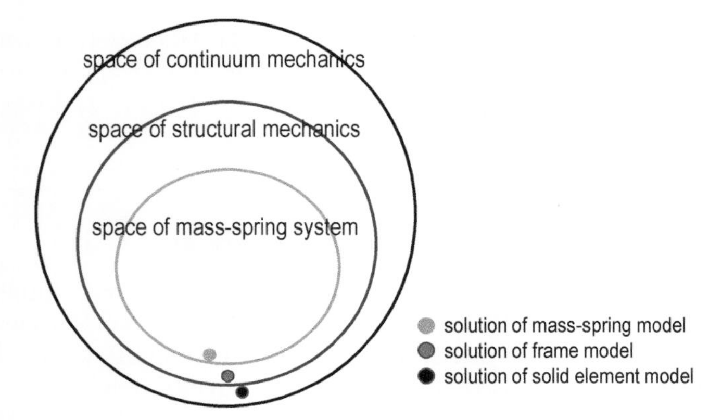

FIGURE 2.1: Function spaces of continuum mechanics, structural mechanics, and mass-spring system.

the mass-spring system are regarded as approximate solutions of this exact solution, even though these two solutions are obtained by exactly solving the governing equations of structural mechanics and the mass-spring system. These approximate solutions are considered to be good approximations if they are close to the exact solution in the function space of continuum mechanics.

2.2.2 Mass-spring model consistent with continuum mechanics model

As an illustrative example, we mention that the beam theory, which solely uses Young's modulus as an elastic property, is rigorously derived from the continuum mechanics of a linear isotropic elastic body that uses both Young's modulus and Poisson's ratio. There is no need to assume a one-dimensional stress-strain relation between the normal strain and stress components via Young's modulus. The governing equation of the beam theory is derived just by making a suitable mathematical approximation, which restricts the forms of displacement and stress function.

In the above derivation, we use a *Lagrangian*, a functional for displacement functions. We consider a structure that consists of a linearly elastic material. Denoting this structure by S, we define the following Lagrangian for its displacement function, $\boldsymbol{u}$:

$$\mathcal{L}[\boldsymbol{u}] = \int_S \frac{1}{2}\rho\dot{\boldsymbol{u}} \cdot \dot{\boldsymbol{u}} - \frac{1}{2}\boldsymbol{\nabla}\boldsymbol{u} : \boldsymbol{c} : \boldsymbol{\nabla}\boldsymbol{u}\,\mathrm{d}v, \tag{2.10}$$

where ρ and $\boldsymbol{c}$ are the density and the elasticity tensor, $\dot{(.)}$ and $\boldsymbol{\nabla}(.)$ are the temporal and spatial derivative of (.), and $\cdot$ and $:$ are the inner product and the second-order contraction, respectively; [] is used for $\mathcal{L}$ to emphasize that $\mathcal{L}$ is a functional. The wave equation is derived from $\delta \int \mathcal{L}\,\mathrm{d}t = 0$. Another equation can be derived from $\mathcal{L}$ when $\boldsymbol{u}$ of a particular form, which is the restriction of a function and regarded as a mathematical approximation, is used for this $\mathcal{L}$. The governing equation of the beam theory is derived in this manner.

We consider another mathematical approximation of $\boldsymbol{u}$ to derive the governing equation of the mass-spring system, paying attention to the mode of structural seismic response. Thus, $\boldsymbol{u}$ is approximated as a sum of the temporal functions of the mode amplitude and the spatial functions for the mode shape, as follows:

$$\boldsymbol{u}(\boldsymbol{x},t) = g(t)\boldsymbol{e} + \sum_{\alpha} U^{\alpha}(t)\boldsymbol{\phi}^{\alpha}(\boldsymbol{x}), \tag{2.11}$$

where g is the input ground motion with $\boldsymbol{e}$ being the unit vector, and U^{α} and $\boldsymbol{\phi}^{\alpha}$ are the amplitude and shape of the α-th mode, respectively. For simplicity, we consider the first mode only and rewrite Eq. (2.11) as $\boldsymbol{u} = g\boldsymbol{e} + U\boldsymbol{\phi}^1$. Substituting this $\boldsymbol{u}$ into Eq. (2.10) yields

$$\mathcal{L}[U] = \frac{1}{2}M(\dot{g} + \dot{U})^2 - \frac{1}{2}Ku,$$

where $M = \int_S \rho\boldsymbol{\phi}^1 \cdot \boldsymbol{\phi}^1 \, \mathrm{d}v$ and $K = \int_S \nabla\boldsymbol{\phi}^1 : \boldsymbol{c} : \nabla\boldsymbol{\phi}^1 \, \mathrm{d}v$; $\boldsymbol{\phi}^1$ is standardized to satisfy $\int_S \rho \, \mathrm{d}v = M$ and assumed to satisfy $\int_S \rho\boldsymbol{e} \cdot \boldsymbol{\phi}^1 \, \mathrm{d}v = M$. As is seen, the above $\mathcal{L}$ is a Lagrangian of the mass-spring model (or a single-degree-of-freedom model), from which the following differential equation is derived:

$$M\ddot{U} + KU = -M\ddot{g}. \tag{2.12}$$

This is the governing equation of the mass-spring model; see Fig. 2.2a. This U is solved for a given g with the initial conditions of $U = 0$ and $\dot{U} = 0$.

The mass-spring model is readily extended to a multi-degree-of-freedom model. For example, we consider the case of two modes, assuming that the

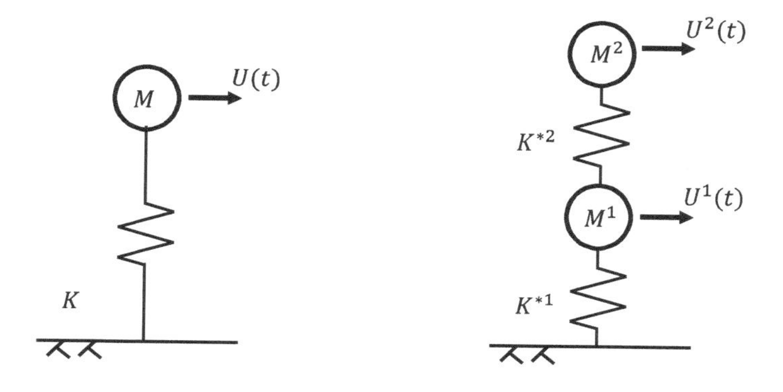

FIGURE 2.2: Mass-spring model and multi-degree-of-freedom model.

two mode shapes, denoted by $\boldsymbol{\phi}^1$ and $\boldsymbol{\phi}^2$, satisfy $\int_S \rho\boldsymbol{\phi}^1 \cdot \boldsymbol{\phi}^2 \, \mathrm{d}v = 0$. The substitution of Eq. (2.11) of the two mode shapes into Eq. (2.10) yields

$$\mathcal{L}(U^\alpha) = \sum_\alpha \frac{1}{2} M^\alpha (\dot{g} + \dot{U}^\alpha)^2 - \sum_{\alpha,\beta} \frac{1}{2} K^{\alpha\beta} U^\alpha U^\beta,$$

where $M^\alpha = \int_S \rho\boldsymbol{\phi}^\alpha \cdot \boldsymbol{\phi}^\alpha \, \mathrm{d}v$, and $K^{\alpha\beta} = \int_S \boldsymbol{\nabla}\boldsymbol{\phi}^\alpha : \boldsymbol{c} : \boldsymbol{\nabla}\boldsymbol{\phi}^\beta \, \mathrm{d}v$; $g\boldsymbol{e} + \sum U^\alpha \boldsymbol{\phi}^\alpha$ is approximated as $\sum (g + U^\alpha)\boldsymbol{\phi}^\alpha$. If we assume that the second term of this $\mathcal{L}$ can be rewritten as

$$\sum_{\alpha,\beta} \frac{1}{2} K^{\alpha\beta} U^\alpha U^\beta \approx \frac{1}{2} K^{*1} (U^1)^2 + \frac{1}{2} K^{*2} (U^2 - U^1)^2,$$

using suitable constants K^{*1} and K^{*2}; then, this $\mathcal{L}$ becomes a Lagrangian of a two-degree-of-freedom model. The governing equation of this model is given as

$$\begin{aligned} M^1 \ddot{U}^1 + K'^1 U^1 - K'^2 (U^2 - U^1) &= -M^1 \ddot{g}, \\ M^2 \ddot{U}^1 + K'^2 (U^2 - U^1) &= -M^2 \ddot{g}. \end{aligned} \tag{2.13}$$

For given ρ and $\boldsymbol{c}$, we can construct a mass-spring model or a multi-degree-of-freedom model by substituting $\boldsymbol{u}$ of Eq. (2.11); the parameters of these two models are computed using ρ and $\boldsymbol{c}$.

The two models explained above provide an approximate solution of the original $\mathcal{L}[\boldsymbol{u}]$, because functions of a restricted form, Eq. (2.11), are used for $\boldsymbol{u}$. The variation of $\mathcal{L}$ with respect to non-restricted $\boldsymbol{u}$ leads to the wave equation,

$$\rho \ddot{\boldsymbol{u}} - \boldsymbol{\nabla} \cdot (\boldsymbol{c} : \boldsymbol{\nabla} \boldsymbol{u}) = \boldsymbol{0}, \tag{2.14}$$

while the variation of $\mathcal{L}$ with respect to the restricted $\boldsymbol{u}$ results in Eqs. (2.12) and (2.13). The calculation of the variation is a mathematical operation, and as explained before, we need not make any physical assumption in deriving Eqs. (2.12) and (2.13) from $\delta \int \mathcal{L} \, \mathrm{d}t = 0$; mathematical approximations are needed, and the accuracy of using the approximations is examined comparing the solution of Eqs. (2.12) and (2.13) with the exact solution of Eq. (2.14).

In order to determine the natural frequencies and related mode shapes of S, we can derive the following eigen-value problem from Eq. (2.14):

$$\omega^2 \rho \boldsymbol{\phi}(\boldsymbol{x}) - \boldsymbol{\nabla} \cdot (\boldsymbol{c} : \boldsymbol{\nabla} \boldsymbol{\phi}(\boldsymbol{x})) = \boldsymbol{0}, \tag{2.15}$$

with suitable homogeneous boundary conditions being posed on ∂S. Here, ω and $\boldsymbol{\phi}$ are the natural frequency and mode shape, respectively. This eigen-value problem has multiple solutions. Denoting the α-th solution by ω^α and $\boldsymbol{\phi}^\alpha$, we can readily prove that $\{\boldsymbol{\phi}^\alpha\}$ satisfy $\int_S \boldsymbol{\phi}^\alpha \cdot \boldsymbol{\phi}^\beta \, \mathrm{d}v = 0$ and $\int_S \boldsymbol{\nabla}\boldsymbol{\phi}^\alpha : \boldsymbol{c} : \boldsymbol{\nabla}\boldsymbol{\phi}^\beta \, \mathrm{d}v = 0$ for $\alpha \neq \beta$. It should be noted that the natural frequency and mode shape obtained by solving Eq. (2.15) do not coincide with those of a multi-degree-of-freedom model due to the mathematical approximations.

As a simple example, we consider the two-degree-of-freedom model, Eq. (2.13). This set of equations has two pairs of an eigen-value and eigen-vector, written as ω^α and $(1, e^\alpha)$, respectively, for the α-th mode ($\alpha = 1, 2$). While the set includes four parameters, $\{M^1, M^2, K^{*1}, K^{*2}\}$, it only contains three independent parameters, say, $\{K^{*1}/M^1, K^*2/M^2, M^1/M^2\}$. Thus, in general, we cannot determine the three independent parameters for the four independent variables that are included in a given set of ω^α and $(1, e^\alpha)$; see Fig. 2.2b. The disagreement of the natural frequency does not mean that the two-degree-of-freedom model should not be used, because with a small difference in the natural frequency, the model can provide a good approximate solution up to the second mode.

2.2.3 Extension of mass-spring model

A Lagrangian of linear isotropic elasticity is readily extended to a nonlinear material. The strain energy density term, $\frac{1}{2}\nabla \boldsymbol{u} : \boldsymbol{c} : \nabla \boldsymbol{u}$, is replaced by a suitable strain energy density term that accounts for nonlinear material properties such as plasticity. We thus construct a nonlinear mass-spring model or a multi-degree-of-freedom model using the Lagrangian and the assumed mode shapes. The nonlinear material properties change the spring constant. Hence, a simpler treatment of the nonlinear case is to replace the linear spring constant, K^1 or $K^{\alpha\beta}$, with a nonlinear spring constant, which changes depending on the amplitude of the mass point displacement. This treatment, however, loses consistency with continuum mechanics since physical approximations are made for determining the nonlinear spring constant. This approximation is inevitable because nonlinear response takes place at every point and the spring constant is determined as the integration of the point-wise response. The relation between the point-wise nonlinear response and the mass point displacement is complicated and cannot be expressed in the simple form of $K = \int_S \nabla \boldsymbol{\phi} : \boldsymbol{c} : \nabla \boldsymbol{\phi} \, \mathrm{d}v$ of the linear material.

The effects of soil-structure interaction are readily included in the Lagrangian of continuum mechanics, just by extending the integral domain of $\mathcal{L}$; if surface soil is denoted by G, the integral domain extends from S to $S + G$ and the material properties of G are used. In an identical manner to that shown above, the effects of soil-structure interaction on the structural seismic response are expressed in terms of a soil spring, i.e., a simplified model such as a mass-spring model with a soil spring or a multi-degree-of-freedom model with soil springs are derived from the variation of $\mathcal{L}$. These simplified models provide an approximate solution of the wave equation, although the partial differential equations are replaced by ordinary differential equations. It should be mentioned that the mass of the soil is omitted in these simplified models. In calculating the variation of $\mathcal{L}$, the soil domain G is separated from the structure domain S, and we do not consider the governing equation in G. The effects of G on S appear in the governing equation in S, taking the form of the soil spring. Appendix B presents a more detailed explanation about formulating a soil spring which is consistent with continuum mechanics; see also Riaz *et al.* (2021).

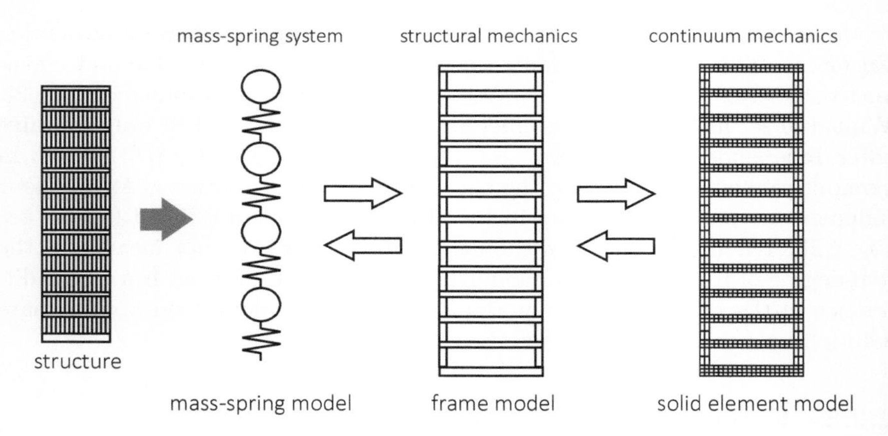

FIGURE 2.3: Meta-modeling of structure for seismic response analysis.

In closing this section, we point out that a more sophisticated model can be used for structural seismic response analysis if digital data of the structures are available. For a given structure, we can construct a mass-spring model according to the mass-spring system, a frame model according to structural mechanics, and a solid element model[2] according to continuum mechanics, which is analyzed by FEM; see Fig. 2.3. The three models provide solutions shown in Fig. 2.1, and the solutions of the mass-spring model and the frame model are a good approximate solution if they are close to the solution of the solid element model. It should be pointed out that if a structure is well designed, the two approximate solutions of the mass-spring model and the frame model are close to the exact solution of the solid element model. It should be noted that the mass-spring model and the frame model must be solved exactly, and the solutions obtained are the exact solution of these models but the approximate solution of the wave equation, i.e., the exact solution of a mathematical problem that is derived from a Lagrangian with restricted functions is an approximate solution of the original variational problem of the Lagrangian. This is the consequence of the meta-modeling theory mentioned before.

2.3 Agent-based simulations of mass evacuation

Time-critical mass evacuations, like those triggered by 2004 Tsumatra Earthquake and 2011 Great East Japan Earthquake, force a large number to

[2]A few programs of FEM enhanced with HPC capability are being developed for structural seismic response analysis; see, for instance, Miyamura *et al.* (2015) and Miyamura *et al.* (2016) for steel structures and Tanaka *et al.* (2016) for reinforced concrete structures.

evacuate within several tens of minutes, often leading to a large number of fatalities. Hence, it is a logical choice to utilize advanced numerical models that can model all the major factors influencing the evacuation progress. Although sophisticated *Agent-Based Models* (ABMs), such as EXODUS and EGRESS2002 (see Kuligowski *et al.* 2005 for a detailed list of ABMs and comparison), are used to simulate fire evacuation from shopping malls, underground stations, stadiums, etc., 1D network flow models and simple ABMs are widely used for simulating mass evacuations such as those triggered by tsunamis. These simple models have several advantages such as the ease of model development, ease of use, low computational demand, and capability of capturing common scenarios such as traffic congestion and pedestrian flow reasonably well.

Although simple ABMs introduce heterogeneity and individual-level actions and interactions, the agents are not fully autonomous. As an example, to ensure that the evacuation time of each agent is realistic, the moving speed of an individual agent is controlled according to suitable *fundamental diagrams* (i.e., speed vs. density relations); see, for instance, Goto *et al.* (2012) and Imamura *et al.* (2012). This lack of fully autonomous agents makes these simple models either ineffective or incapable of modeling scenarios such as pedestrians and cars interacting on roads, interruptions to traffic and pedestrians due to debris on road, vehicles and pedestrians at non-signalized junctions, and visibility restriction. Though extremely rare in daily life, these scenarios are not only common but also significantly impede the progress of evacuation during a major natural disaster such as a mega tsunami. A general and versatile approach is to develop ABMs with fully autonomous agents capable of reproducing realistic outcomes as emergent phenomena from the actions and interactions of individual agents (Wang *et al.* 2021). Once autonomous software agents with sufficient capabilities to mimic real evacuees' behaviors are developed, ABMs can be utilized to model different combinations of these rare scenarios. As an example, if algorithms are developed to make autonomous agents reproduce fundamental characteristics such as the moving speed in any dense crowd as an emergent phenomenon, it will be able to reproduce a large array of the real-world phenomenon involving pedestrians and vehicles to a reasonable accuracy. Wang *et al.* (2021) and Mas *et al.* (2015) provide detailed accounts on the application of autonomous agents for tsunami evacuation. Several publications have proposed ABMs consisting of sophisticated agents capable of reproducing fundamental diagrams of continuous 2D space pedestrian movements (Curtis *et al.* 2014, Sahil *et al.* 2015, and Makinoshima *et al.* 2018) and different models of vehicles (Gao 2011). The interactions of different modes of evacuation have been studied in non-continuous 2D microscopic models (Goto *et al.* 2012, Xin *et al.* 2014, and Jiang *et al.* 2006). It is rare to find vehicle pedestrian interaction models at non-signalized junctions, and even the existing vehicle-vehicle interaction models, which are mostly based on collision avoidance algorithms (Javier *et al.* 2015, Berg *et al.* 2011, Campos *et al.* 2013, Doniec *et al.* 2008a, Doniec *et al.* 2008b, and Fu *et al.*

2016), scheduling schemes (Colombo *et al.* 2015), game theory (Rene *et al.* 2008), etc., do not reproduce realistic trajectories and speed profiles. While the development of the sophisticated agents capable of producing real-life observations as emergent phenomena itself is challenging, the need for scalable high-performance computer implementations to meet the computational demand of sophisticated agents further increases the required efforts. Though several parallel computing extensions have been implemented to meet the computational demand of a large number of sophisticated agents, these implementations are low in parallel efficiency and most can accommodate only several ten thousands of agents (see Makinoshima *et al.* 2018 for a detailed account available for parallel implementations).

With the aim of simulating scenarios influential in emergency mass evacuations, we developed an ABM with sophisticated autonomous agents and a high-resolution hybrid model of the environment. The developed agents are capable of perceiving the features of the high-resolution environment, storing past experiences, updating the decisions like a route taking the past experiences in to account, reproducing speed-density characteristics as the emergent behavior, etc. (see Leonel *et al.* 2019, Leonel *et al.* 2016, Leonel *et al.* 2014, Leonel *et al.* 2013, and Leonel *et al.* 2014). In order to meet the high computational demand of these sophisticated agents, we implemented a High-Performance Computing (HPC) extension with a reasonably high parallel computing scalability (Leonel *et al.* 2017, Wijerathne *et al.* 2018, and Wijerathne *et al.* 2013). While our HPC-enhanced evacuation simulator is designed to work on any modern computer ranging from laptop-PCs to supercomputers, our largest simulation with 10 million agents in 588 km^2 scaled up to 2048 CPUs (16384 CPU cores) in the K supercomputer. The rest of this subsection provides an abstract model to give an overview of the developed ABM, some of the algorithms, and a brief introduction to the HPC extension.

The developed ABM is a time-step driven (i.e., the agent-based model advances at equal time steps of Δt and all the agents share a common clock) dynamic system consisting of a high-resolution grid environment that updates according to the physical disaster being simulated, a topological graph that represents an abstract model of the traversable space of the grid, and autonomous agents mimicking the evacuees. The purpose of the grid is to include a sufficiently accurate model of the environment, and does not influence the minimum distance an agent can move within a Δt period. An agent can move to any location inside a grid cell according to its current location, chosen moving speed, and direction; agents do not hop from cell-to-cell like in cellular automata.

2.3.1 Mathematical framework

Let the state of the simulated physical domain (i.e., the environment in ABM terminology) at time t be represented by $E(t)$. The discrete time evolution of the environment can be expressed as $E(t + \Delta t) = \Lambda(E(t), t)$, where Λ is an

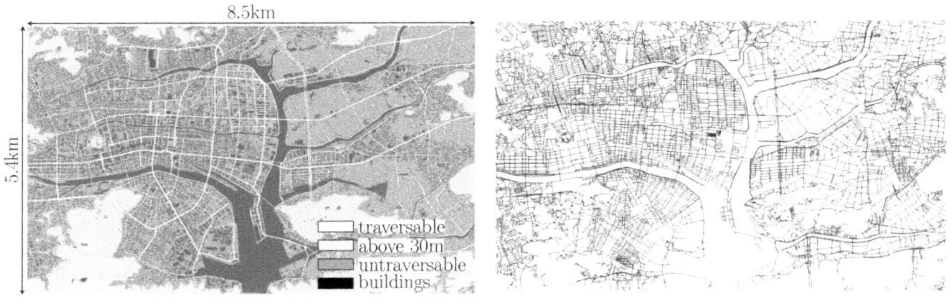

a) 1 m×1 m resolution grid. b) topological graph

FIGURE 2.4: Grid and topological graph of Kochi city area.

external function, which updates the environment according to the physical disaster under consideration (Fig. 2.4).

Let $A = \{a_i | i = 1, \dots, n\}$ be the set of n agents. The ith agent is defined as $a_i = \{s_i, f_i\}$, where s_i and f_i represent the agent's state and the possible actions, respectively. The state consists of private and public parts, $s_i = \{s_i^{prv}, s_i^{pub}\}$. The public part s_i^{pub}, which can be observed or deduced by fellow evacuees, consists of parameters such as speed, walking direction, size, gender, and age category. The rest of the information, which is not deducible by or inaccessible to the fellow agent unless communicated, like his past experiences, the blocked roads it encountered, its target destination and route, etc., compose the private part s_i^{prv}; $s_i^{prv} \cap s_i^{pub} = \emptyset$. Similarly, the actions consist of the public part f_i^{pbl} and the private part f_i^{prv} (i.e., $f_i = \{f_i^{pbl}, f_i^{prv}\}$). f_i^{pbl} includes the function to exchange information with the neighboring agents, while all the rest, such as decision-making and moving, compose the private part f_i^{prv}[3]. Let $E_i^{vis}(t) \subset E(t)$ be the region of the environment visible to the agent a_i at time t, and the public states of all the agents visible to a_i be $\bar{S}_i = \{\bar{s}_j^{pub} | i \neq j\}$. Then, the discrete time evolution of a_i can be expressed as

$$s_i(t + \Delta t) = f_i(s_i(t), E_i^{vis}(t), \bar{S}_i(t), t).$$

Note that the actions of an agent are designed to depend only on the states at the current time t so that the time evolution of the agents is independent of the execution sequence of the agents; hence, the results from HPC implementations (i.e., the parallel execution of the agents) are independent of the utilized number of threads and computing cores.

[3]In object-oriented programming languages such as `C++`, agents can be implemented as self-contained objects, and their private and public states and functions can be implemented as `public` and `private` members of the corresponding classes.

2.3.2 Hybrid model of the environment

The environment is modeled as a hybrid of a grid G and a graph $\mathcal{G}$ (Fig. 2.4) so that the strengths of each of these data structures can be exploited to efficiently simulate hundreds of square kilometers size domain in high resolution. The state of the physical environment under consideration is approximated as a two-dimensional (2D) Cartesian grid $G^t = \{g_{ij}(t)|i, j \in \mathbb{N}\}$, where $g_{ij}(t)$ defines the state of the grid cell (i, j) at time t. The discrete time evolution of the environment is defined as $\Lambda : G^t \mapsto G^{t+m\Delta t}$, where Λ is an external function with which the state of each cell $g_{ij}(t)$ is updated at suitable time intervals of $m\Delta t$ $(m \in \mathbb{N})$, according to the time evolution of the physical disaster (e.g., the progress of tsunami inundation and the scattering of debris). The physical environment can be modeled to a reasonable accuracy by choosing a sufficiently high spatial grid resolution (e.g., $1\,\text{m}\times 1\,\text{m}$) and assigning distinct integer values to each cell according to the possible states it can take (e.g., $g_{ij} = -1$ if an evacuation area, 0 if traversable, 1 if non-traversable, 2 if occupied by water, etc.). The agents are equipped with the functions to scan their visible grid neighborhood in high resolution and identify the features to mimic the evacuees' visual perception of the surrounding, providing a base for sophisticated sensing and behaviors.

The main downsides of using a high-resolution grid are that path planning on very large grids is time-consuming, and storing agents experiences with reference to the grid cells is memory consuming. In order to eliminate these problems, connectivity of the traversable spaces in the grid is abstracted by a topological graph $\mathcal{G} = \mathcal{G}(\mathcal{L}, \mathcal{V})$, where $\mathcal{L} = \{l_i|i \in \mathbb{N}\}$ and $\mathcal{V} = \{v_i|i \in \mathbb{N}\}$ are the sets of links and nodes defining the bi-directional graph. The links contain physical characteristics like the width and length of the road segment, number of vehicle lanes and allowed flow directions, etc. Even though the grid is updated according to the physical disaster, neither the connectivity of the graph nor the physical characteristics stored in each link are updated. Therefore, the graph always contains topological connectivity of an ordinary day, severing as the base map in agents' decision-making.

The graph provides support for agents' thought processes, such as path planning, otherwise computationally intensive with a grid environment. Scanning the grid in high resolution, agent a_i recognizes the features and changes of its visible surrounding, and important changes, like blocked or inundated roads, etc., and stores in its private state s_i^{prv} with reference to the links and nodes of the graph. The graph is equipped with various path planning functions to support agents' decision-making process. An agent can include its past experiences in to decision-making by sending the desired information, like the list of blocked paths it encountered, to these functions. Further, statistics like the current number of agents at a certain stretch of a road, number of vehicles and pedestrians passed, etc. are collected to the corresponding links of the graph at desired time intervals. These statistics are useful for various purposes like making the agents avoid crowded routes (Leonel *et al.* 2016),

Algorithm 3: Pseudo code for width-preferred path planning algorithm accepting functors with custom data and conditions.

```
input  : Starting node n_s, destination node n_e, preferred width w_pref,
         desired minimum width w_min, desired maximum distance l_max,
         functor F
output: path to destination satisfying the conditions of F, and the length
         of the path
// l_max can be set to reflect the need to reach a shelter
within a given time, etc.
// d-distance,pd-perceived distance, nd-new distance, npd-new
perceived distance
// the two member functions l(u,v) and w(u,v) of the graph
return the length and width of the link from node u to v
for ∀u ∈ V do {pd(u)= ∞; parent(u)= −1;} ;
pd(n_s)= 0; d(n_s)= 0;
priority_queue.add(n_s,pd(n_s));
while not priority_queue.empty() do
    u = priority_queue.top();
    if u == n_e then break();
    priority_queue.pop();
    if F (u) then                          // can pass through node u?
        for ∀(u, v) ∈ E do
            if F (u,v) then        // can pass along link from u to v?
                if w (u,v)< w_min then Δpl = l(u, v) ;
                else if w (u, v)<= w_pref then Δ = l (u,v) / w (u,v) ;
                else Δpl = l (u,v) / w_prf ;
                npl = pd(u) +Δpl;
                nd = d(u) + l (u, v);
                if (pdu, v > npl) and (nd < l_max) then
                    pdv = npd; v = nd; parent(v) = u;
                    priority_queue.add(v,pdv);
path.clear();
if u == n_e then                             // Extract the path to n_e
    length = d(u); path.pushBack(u);
    while u ≠ n_s do { u = parent(u); path.pushBack(u); } ;
return path , length
```

Algorithm 3: Contd.

```
// A customizable functor with operators to check whether a
// given node u or a link from u to v is available to traverse

struct {
    // set the nodes and links to exclude based on each agent's
    experiences;
    exclude_node_list, exclude_link_list, etc.;
    operator (u) // Update the logic as necessary
1       if (u ∈ exclude_node_list) then return false;
2       return true;
    operator (u, v) // Update the logic as necessary
3       if ((u, v) ∈ exclude_link_list) then return false;
4       return true;
}F;
```

analyzing the progress of evacuation and identifying bottlenecks, passing the statistics of one simulations to another in optimization problems, etc.

In the implementation, most of the algorithms included in the graph are `C++ template` functions which accept `functors` inside which custom data and functions can be embedded, thereby facilitating a wide array of path planning and other supportive functions. As an example, the algorithm 3 shows a pseudo code of modified Dijkstra's algorithm (Dijkstra 1959) for minimizing the use of narrow roads in night-time evacuation scenarios, while avoiding any undesired list of nodes and links according to an agent's past experiences (Leonel *et al.* 2014 and Leonel *et al.* 2016). Figure 2.5 shows some examples of paths found using the algorithm 3. Further, cell states of G are defined by single byte integers (e.g., `std::int8_t` in C/C++11) in order to reduce the memory usage (i.e., $\{g_{ij} | g_{ij} \in \mathbb{Z}, -127 \geq g_{ij} \geq 127\}$).

2.3.3 Agents

Heterogeneity in crowds is introduced by varying each agent's state, s, and the local update function f. As it would be impractical to specialize every f_i and s_i mutually exclusive subsets, $F^\tau \subseteq F$, $S^\tau \subseteq S$ are created, where $F = \{f_1, .., f_n\}$ and $S = \{s_1, ..., s_n\}$. τ stands for the agent type label. From now on $a^\tau = \{f^\tau, s^\tau\}$ will refer to a representative member of the specialized subgroup with label τ. f^τ is specialized based on the role of an agent and the information it possesses, while s^τ is specialized based on agent's physical capabilities and role.

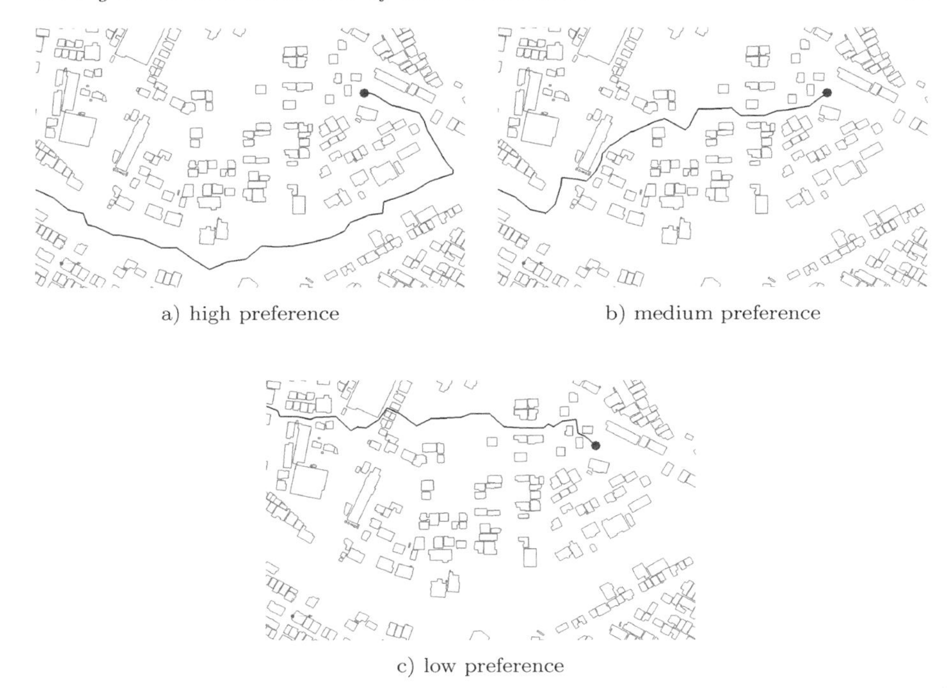

FIGURE 2.5: Some examples of paths with preferred widths found using algorithm 3.

The general local update function – f

Potentials of an agent-based model mainly depend on the agent's update functions f, which encompass an agent's behavior, actions and interactions models. f is composed of a basic set of constituent functions g^i as $f = g^1 \circ g^2 \circ ... \circ g^m$. Some of the implemented constitutive functions are briefly explained below.

g^{see}**:** Scans the grid G and creates the boundary of visibility (see Fig. 2.6) in s^{int} based on agent's sight distance, which can be 50 m or longer depending on an agent's physical abilities.

$g^{identify\_env}$**:** Analyze visual boundary and extracts features such as open paths.

$g^{navigate}$**:** Chooses a suitable open path based on the obstacles and openings identified in $g^{identify\_env}$ and the route followed by the agent.

$g^{identify\_inter}$**:** Recognizes neighbor agents to interact with, based on visibility, interaction radius, etc.

g^{coll_av}**:** Finds a collision-free walking direction along which an agent can move closer to its preferred speed for a given minimum period of comfortable time

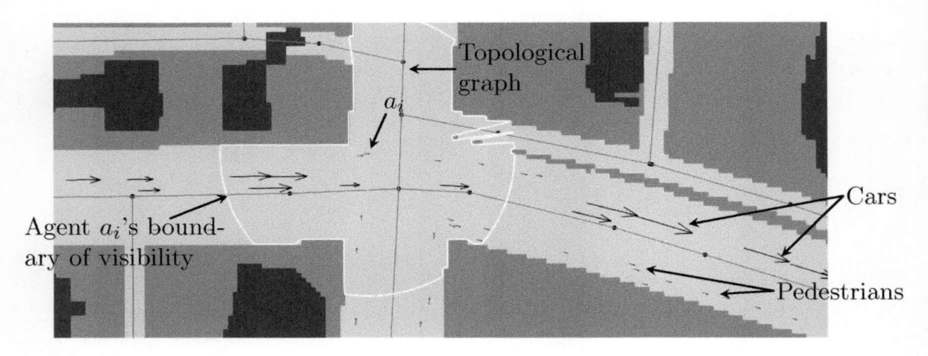

FIGURE 2.6: Snapshot of agents' movements at a junction. Gray and black arrows indicate instantaneous velocities of pedestrians and cars, respectively. Pedestrian agents walk along the edges, if the road can accommodate vehicles.

$t^{comfort}$ to reach the opening chosen in $g^{navigate}$, evading collision with the neighbors identified in agents $g^{identify\_inter}$. The preferred comfortable time $t^{comfort}$ and other parameters are tuned to reproduce observed fundamental diagrams.

$g^{path\_planning}$: Finds paths with desired characteristics.

$g^{is\_path\_blocked}$: When navigating in damaged environment, analyzed the data from g^{see} to identify whether the desired path is blocked.

$g^{find\_and\_follow}$: Finds a suitable agent and follows it, if the environment is not familiar.

g^{side_walk}: Use pedestrian sidewalks, when available

$g^{car\_at\_signalized\_junction}$: Control the movement of car agents at junctions obeying traffic signals, maintaining safe distances with cars and pedestrians on potential collision courses, and maintaining safe speeds according to the curvature of trajectory.

$g^{car\_at\_unsignalized\_junction}$: Extended version of $g^{car\_at\_signalized\_junction}$ without centralized flow control by traffic lights to enable cars move at unsignalized junctions.

$g^{\cdots}$: etc.

$g^{execute\_actions}$: Executes desired actions such as move

g^{update}: Updates an agent's state

The above constituent functions can be grouped into three main subsets as *See*, *Think* and *Act*. *See* consists of g^{eyes}. *Think* encompasses the exploratory behavioral scenarios providing a basic workbench for an agent designer to compose desired behavioral models using subsets of predefined constituent functions like $g^{identify\_env}$, $g^{navigate}$, g^{coll_av}, etc., or adding any desired. *Act* is composed of $g^{execute\_actions}$ and g^{update}.

Agents specializations

Agent types, denoted by a^{τ}, with different behaviors are introduced by specializing *Think*. As examples, short descriptions of four types of implemented agents are given below.

$a^{resident}$: Represents a local resident of the simulated area. $s^{resident}$ possess a mental map of the environment (i.e., access to $\mathcal{G}$ and most path finding algorithms), and uses $g^{path\_planning}$ to find paths according to its desired constrains and past experiences. Additionally they know the location of possible evacuation areas.

$a^{visitor}$: Represents non-resident people in the interest area. They do not possess any additional information of the environment aside what they can visually perceive (i.e., g^{see}). Its main evacuation mechanism is to seek a visible high ground or follow other evacuees using $g^{find\_and\_follow}$.

a^{car} Represents multiple people (one or more evacuees) traveling by a car. s^{car} possesses a mental map of the environment, and equipped with logic to drive through unsignalized junctions avoiding collisions with pedestrians and other cars.

$a^{official}$: This type of agents represents figures of authority, such as law enforcement, event staff, etc. Their main task is to facilitate fast and smooth evacuation by independently or collectively planning the areas to be covered by each with $g^{path\_planning}$, and commanding or delivering information to other agents with $g^{deliver\_message}$. $s^{officials}$ also possess a mental map of the environment which can be updated through communication.

Figure 2.6 shows a snapshot of car and pedestrian agents at a junctions. The white color partially circular boundary shows the boundary of visibility produced by the constituent function g^{see} of a pedestrian agent. Each individual agents scan its surrounding in a similar manner to identify available paths, obstacles and visible neighboring agents. Agents store obstacle or important experiences in their memory with respect to the corresponding link or node of the graph $\mathcal{G}$ so that those experiences can be taken into account in future decision-making.

2.3.4 Validation of constituent functions

Validation of g^{coll_av}

The constituent functions for collision avoidance, g^{coll_av}, play a central role in the ABM's ability to reproduce the fundamental characteristics (i.e., speed vs density relation of agents) as emergent behaviors. Our collision avoidance algorithms are developed based on the Optimal Reciprocal Collision Avoidance (ORCA) (Berg *et al.* 2011) algorithm that construct a convex polygon of possible velocities at which an agent can move for a desired short period avoiding collision with its neighbors, and find the optimal velocity using integer programming. In forming convex polygons of feasible velocities, ORCA algorithm eliminates some of the admissible and even the optimal velocities, sometimes leading to deadlocks. We introduced further flexibility to ORCA algorithm by adding freedom to make the agents respond to different crowd densities, and introduced modifications to avoid potential deadlocks and non-optimal group behavior of the original ORCA algorithm (Leonel *et al.* 2013 and Leonel *et al.* 2014). We specialized and tune the parameters to model the three main collisions encountered in the target simulations (Leonel *et al.* 2013 and Leonel *et al.* 2014); $g_{pp}^{coll\_av}$ for pedestrian-pedestrian collision, $g_{cc}^{coll\_av}$ for car-car collision, and $g_{cp}^{coll\_av}$ car-pedestrian collision.

Each of the three specialized algorithms is validated by comparing a selected set of real-life observations with numerical results from the simulations with the settings of the corresponding observation conditions. The constituent function for pedestrian-pedestrian collision avoidance model $g_{pp}^{coll\_av}$ is validated by comparing with field observations by Mori *et al.* (1987), and the observations compiled by Weidmann (1993). As seen in Fig. 2.7a, the simulation results presented as a whisker-box plot are in good agreement with the observations compiled by Weidmann (1993) even up to higher densities. To validate $g_{cc}^{coll\_av}$, we used the observation of Lincoln tunnel reported by Dhingra *et al.* (2008). Their observation conditions are simulated and the speeds and densities of cars are estimated using the same measures used by Dhingra *et al.* (2008).

As seen in Fig. 2.7b, the simulation results are in reasonable agreement with the observations. In both the above cases, published literature provides only the regression data of the observations. Unavailability of original observations or statistical measures like standard deviations is a major problem we encountered in better tuning the model parameters and conducting more detailed comparison. In contrast to the above validations there is no available observation data for validating the car-pedestrian collision avoidance. Though Kwon *et al.* (1998) present observations useful for the modeling of different modes of evacuees' behaviors, no quantitative data usable for validation is available. Because of this, we performed field observations at two locations near Nezu shrine, Tokyo, Japan, during a golden week when the many people visiting the shrine interact with cars on a narrow road. Videos of 72 cars passing through crowds are recorded with an ordinary video camera and the recorded car movements through the crowd were analyzed. Figure 2.7c shows that both the simulation results and our observations follow a similar trend with a similar dispersion, indicating that those are in a reasonable agreement. Further details of these constituent functions for collision avoidance and their parameter tuning are presented in Appendix B.

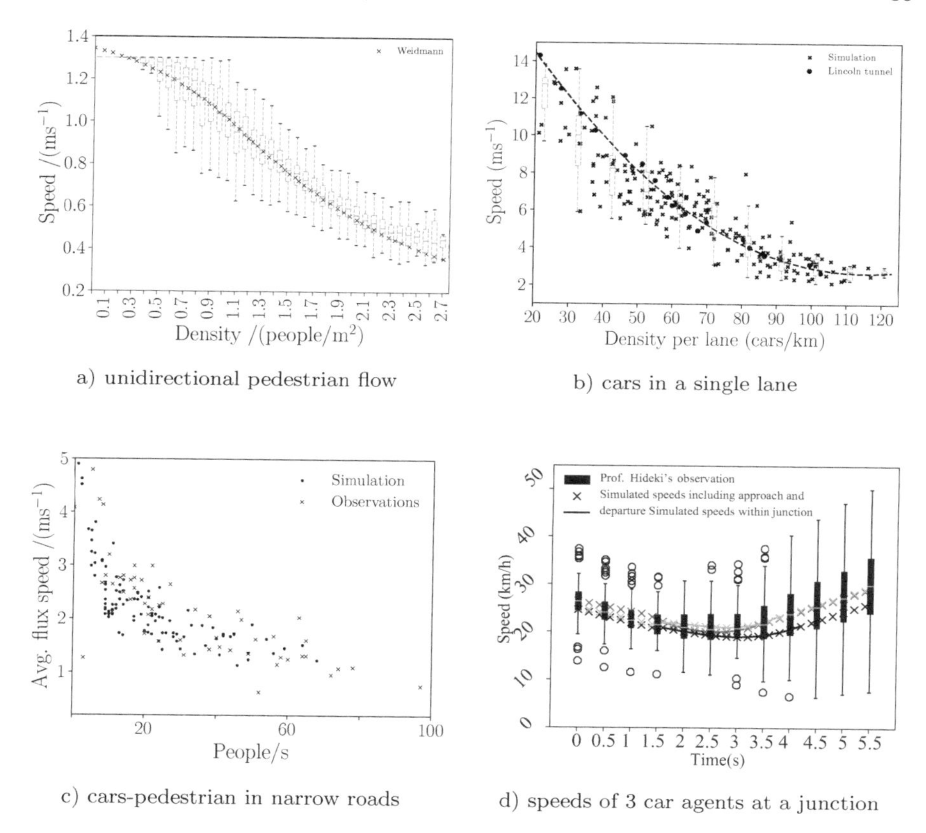

a) unidirectional pedestrian flow

b) cars in a single lane

c) cars-pedestrian in narrow roads

d) speeds of 3 car agents at a junction

FIGURE 2.7: Comparisons of fundamental diagrams from simulations and field observations, and comparison of car agents' speed at junctions with observations.

Validation of speed profiles of car agents at junctions

As later explained in Section 3.3.2, the trajectories of car agents at junctions are approximated using B-splines to accurately mimic the real life. Further, we approximate the free flow speed[4] s_{ff} of a car agent along these curved trajectories using cubic polynomials. According to Dias *et al.* (2017), speed profiles of vehicles at junctions can be approximated with fifth-order polynomials. For the sake of easily identifying the required parameters, we use the following third-order polynomial approximation that can be defined with three known parameters and one constraint.

$$s_{ff}(r) = 4(s_d - s_a)r^3 + 4(2s_a - s_m - s_d)r^2 + (4s_m - 5s_a + s_d)r + s_a, \quad (2.16)$$

[4]Preferred speed of a driver when there is no danger of collision with other vehicles or pedestrians.

where r $(0 \geq r \geq 1)$ is the fraction of the distance traveled along a curved trajectory of length L. As for the constraint, we assumed that the acceleration is zero (i.e., $\frac{ds_{ff}}{dt} = \frac{ds_{ff}}{dr} s_{ff} = 0$) at the point of the highest curvature at where a car reaches the minimum speed of s_m; s_m can be considered as the maximum allowable speed to prevent accidents due to centripetal force. $s_a (= s_{ff}(0))$ and s_d $(= s_{ff}(1))$ are the approaching speed and the desired departing speed. Most trajectories have the maximum curvature at distances close to $\frac{L}{2}$, for which the above equation is derived; one can easily derive it for general cases. Further, we randomly set s_a, s_m, and s_d to agree with the field observations, shown as a whisker-box plot in Fig. 2.7d, provided by Prof. Hideki Nakamura, Nagoya University. We found that this simple approximation can reasonably reproduce the observed speed profiles of cars at junctions as shown in Fig. 2.7d; lines indicate the speeds within the junctions, and $\times$ marks include the deceleration and acceleration during the approach and departure. Equation 2.16 defines only the free flow speed, and if a car agent detects potential collision with other cars or pedestrians, it decelerates to a speed $s < s_{ff}$ to prevent collision, and accelerates back to s_{ff} once it is clear of any collisions.

2.3.5 HPC extension

To meet the high computational demand of simulating large number of autonomous agents spread over a large domain, a scalable High-Performance Computing (HPC) extension was developed. Most of the agents' constituent functions are both computation and memory intensive. As examples, g^{see} scans the grid in high resolution within an agent's sight distance using a computation and memory intensive ray tracing algorithm (Leonel *et al.* 2017), $g^{is\_path\_blocked}$ scans the grid to check whether an agent can navigate through a debris field to reach its destination, etc. Consequently, simulations demands significant computational resources. As an example, simulation of a 50-minute long evacuation with 90,000 agents in the environment shown in Fig. 2.4 at $\Delta t = 0.2$s required 33 node-hours in the K-computer; a computing node of K-computer consists of an 8-core SPARC64 VIIIfx processor with 16GB of RAM. To meet this high computational demand, a hybrid parallel extension was developed (Leonel *et al.* 2017, Wijerathne *et al.* 2018, and Wijerathne *et al.* 2013). This section briefly presents some details of the implemented HPC extension.

The agents are recursively partitioned into 2N number of geographically continuous rectangular subsets using kd-tree (Bently 1975) such that each partition has nearly the same computational workload. We use the measured execution time of each agent in generating kd-tree to assign nearly equal workload to each partition since individual agent's execution time depends on its type, visible surrounding, agent density it its neighborhood, etc. Each partition is assigned to an MPI process, which is a shared memory compute node in our case, and the computations within a node are accelerated using OpenMP threads. Most of the algorithms of our interacting agents have extensive race

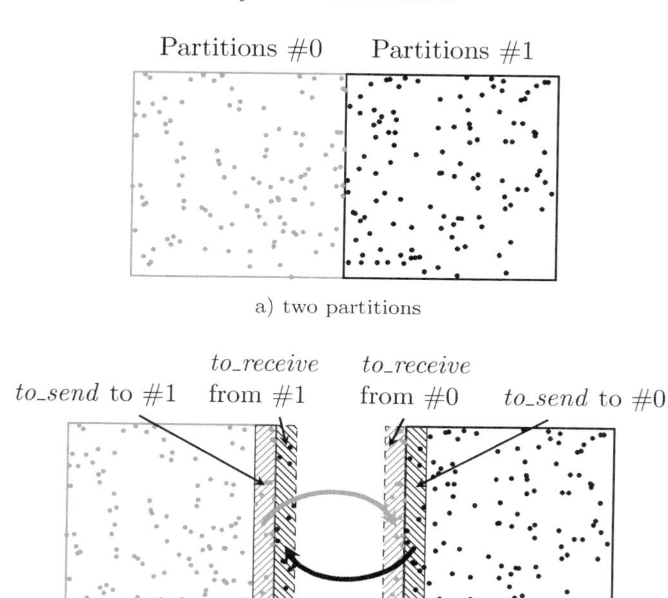

FIGURE 2.8: Illustration of communication between MPI ranks to maintain continuity using ghost updates.

conditions making it difficult to exploit shared memory parallelism. We use a data-redundant approach to avoid most of the race conditions and OpenMP's task-level parallelism to accelerate computations using shared memory parallelism. Since a thread in one compute node cannot access the memory of the compute nodes holding neighboring partitions, we maintain overlapped boundary regions, which are usually called ghost regions or halos, as shown in Fig. 2.8. The states of the agents lying in ghost regions are exchanged using MPI at each time step to maintain the continuity of the simulation over multiple compute nodes. Since agents' behaviors are influenced by their neighbors within their sight distance, the width of the ghost region is set to be larger than the longest of the agents' eyesight.

Communication hiding

MPI communications for maintaining the continuity among nodes is an extra overhead, which increases with the number of partitions. Performing some useful computation while communications are happening in the background, which is known as communication hiding, is a widely used standard technique to minimize the performance degradation due to such communication overheads. To facilitate communication hiding, we group the agents in an MPI process into three mutually exclusive sets as shown in Fig. 2.8: *to_send* includes all the agents to be sent to neighboring MPI processes, *to_receive* includes all the agents in ghost regions, while the rest are included in *inner_most*. We hide the communication overhead by first executing the agents in *to_send* and posting non-blocking MPI calls to send those updated data to corresponding neighboring MPI processes and receive in *to_receive* memory regions. While the posted non-blocking communications progress in the background, the agents in *inner_most* region are executed. If execution of *inner_most* takes longer than the time to complete the communication, we can effectively eliminate the communication overhead.

Dynamic load balancing

The movements of agents make them move between the sub-regions *inner_most* and *to_send* forcing to update the list of agents in these sub-regions at each time step. Further, movements of agents from *to_send* to *to_receive* forces us to transfer the ownership of those agents to corresponding neighbor partitions. This permanent transfer of agents between partitions, which we refer as *migration*, is time-consuming. At the same time, the workloads assigned to an MPI process increases/decreases depending on the number of agents migrated in/out to/from a partition. Large differences in workloads introduced by migrations can significantly degenerate the parallel computing efficiency. When the load imbalance among MPI processes reaches a critical state, domain is re-partitioned to re-assign equal workloads to MPI processes. We measure the total execution time of the agents in each MPI process and re-partitioning is called when the time for re-partitioning process is smaller than the computational time wasted due to load imbalance.

Strategies like the data-redundant approach for eliminating data races, communication hiding, combining communication of multiple data with user defined MPI data types, dynamic load balancing, etc., enabled us to implement a scalable MPI-OpenMP hybrid parallel computing code with a high strong parallel scalability. Our simulation of a hypothetical evacuation event involving 10 million agents in 588 km^2 area of central Tokyo in $1\,\mathrm{m}\times 1\,\mathrm{m}$ resolution produced 82% strong parallel scalability with 2048 nodes (2048×8 CPU cores) of the K supercomputer (Leonel *et al.* 2017).

3

Automated Model Construction

CONTENTS

DOI: 10.1201/9781003149798-3

3.1 Underground structures

Linear wave propagation in the earth's crust and nonlinear wave propagation in the soft soil are strongly affected by the geometry of the earth's crust and the soil structure, and in particular, by the stress-free boundary condition of the ground surface. Therefore, in this section, we consider the mesh generation of finite element models for low-order unstructured finite element analysis that excels in modeling the geometry and analytically satisfies the stress-free boundary condition of the ground surface. In particular, when analyzing linear wave propagation in the earth's crust and nonlinear wave propagation in soft soil, the earth's crust and ground are often modeled as stratified structures. Thus, a method for generating unstructured tetrahedral finite element models for stratified structures is considered.

Since the quality of the analysis results of the finite element method depends on the finite element model, various methods such as Delaunay decomposition (Caendish Field and Frey 1985) and advancing front method (Lo 1985) have been proposed, and many methods have been developed to date for generating more efficient and high-performance finite element models. Because the crust and the ground to be analyzed have large areas and require high resolution, the degrees of freedom of finite element models are often huge, making it difficult to construct large-scale finite element models using conventional finite element model generation methods. Therefore, it is necessary to not only increase the scale and speed of the analysis method but also to study the scale of the finite element model generation method. Furthermore, while tetrahedral elements are effective in generating finite element models of arbitrary shapes, the results are greatly affected by their aspect ratio; thus, it is necessary to pay attention to this aspect ratio in order to guarantee the quality of the analysis results. (Ichimura Hori and Bielak 2009) combines the structured background cell with the octree structure to enable the construction of high-quality large-scale finite element models. In this section, we present an automatic finite element model generation method by (Ichimura Hori and Bielak 2009) building a three-dimensional finite element model using tetrahedral elements. In this method, a three-dimensional finite element model is automatically constructed from a stratified structure model (layer boundary data and layer properties). In Step 1, a background cell consisting of a structural grid is placed over the layer boundary data, and in Step 2, the grid containing the layer boundary is divided into tetrahedral elements. In Step 3, the cells away from the layer boundary are integrated by the octree structure. The partitioning of the tetrahedral elements in the background cell can be performed independently, which is suitable for parallel computation, and the number of nodes in the entire model can be reduced using the octree structure. This method can be used on shared memory computers to generate finite element models robustly and quickly. By interpolating the generated

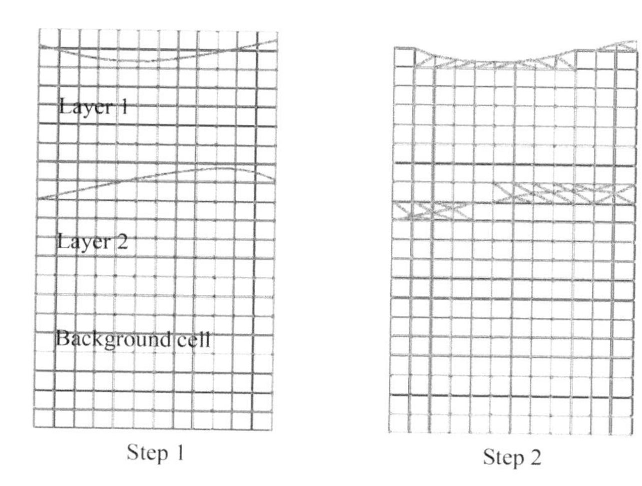

FIGURE 3.1: A method to automatically generate finite element models with unstructured tetrahedral and voxel element.

tetrahedral elements, a large-scale finite element model based on tetrahedral second-order elements can also be constructed. The details of the method are shown below. Note that α is approximately $1/8 \sim 1/10$ for tetrahedral first-order and $1/4 \sim 1/5$ for tetrahedral second-order elements.

Step 1 A structured grid with a spatial resolution ds corresponding to the target frequency (about α times the wavelength of the target frequency) is used as a background cell to cover the target region. The background cells cover the entire 3D model, and the number of cells is determined according to ds. The created background cells and the elevation of each layer are shown in Fig. 3.1.

Step 2 The generated cells that do not contain a ground surface or boundary surface are assumed to be cubic elements. If a cell contains a ground surface or a boundary surface, the following processing is added. If the cell intersects the ground surface or the boundary surface, the cell has at most four edges that intersect the ground surface or the boundary surface. The four intersections on the cell edges and the eight vertices of the cell are used to generate a tetrahedral element by Delaunay decomposition. This method produces elements that are consistent across the cells in a grid-by-grid manner. The advantage of this method is that each cell can be partitioned independently without considering the partitioning of other cells. In the case of element partitioning using Delaunay decomposition, if the distance between the four intersections on an edge and the eight nearest vertices from each intersection is small, an element with a large aspect ratio may be generated. Therefore, a small threshold value is set for the distance between the intersections and

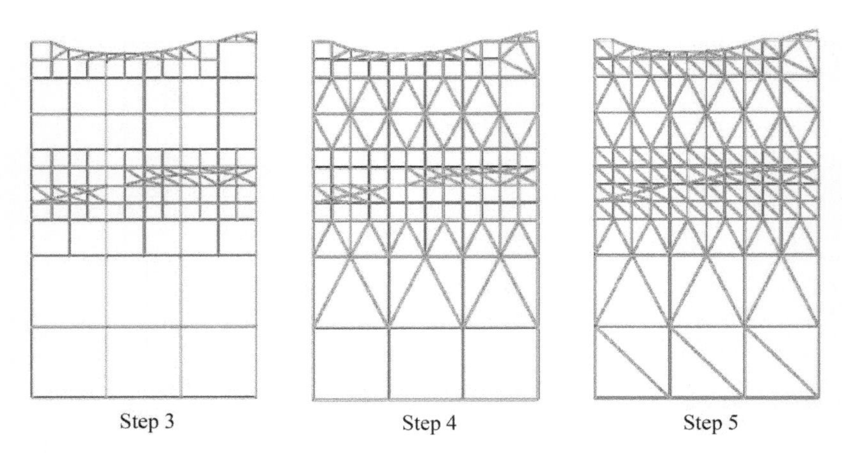

FIGURE 3.2: A method to automatically generate efficient finite element models with the octree structure.

the vertices, and if the distance is smaller than the threshold value, a small approximation of the shape of the ground surface and boundary surface is performed to reduce the approximation error. As described above, tetrahedral and cubic elements are generated in a grid-by-grid manner based on the approximated ground surface/boundary surface and the background structured cell. The approximated surface and boundary, and the generated tetrahedral and cubic elements are shown in Fig. 3.1.

Step 3 In the part of the cubic element where the material properties are homogeneous in the stratified structure, the size of the cubic element can be increased. Therefore, we use the octree structure to increase the size of the cubic element by a factor of ν, with the goal of reducing the number of degrees of freedom to the extent that the numerical solution converges. The ν is a power of 2 (($\nu = 2^i$), where i is a natural number), and the size should be large enough to satisfy the convergence of the solution (the element size should be about α times the wavelength of the target frequency) (see Fig. 3.2). By setting ν independently for each layer, it is possible to generate finer elements in the layers that are to be nonlinearized and coarser elements in the layers that remain linear. This makes it possible to generate more efficient 3D finite element ground models for nonlinear wave field calculations.

Step 4 Since the elements are inconsistent around the cubic elements that have changed size, a tetrahedral element is generated by Delaunay decomposition to connect cubic elements of different sizes (see Fig. 3.2).

Step 5 The generated cubic element is divided into tetrahedral elements by Delaunay decomposition (see Fig. 3.2).

When a high-quality finite element model can be generated, large-scale ground motion analysis can be performed using a matrix solver, which has been extended to handle a large number of degrees of freedom with high-performance computing (e.g., Ichimura *et al.* 2014 and Ichimura *et al.* 2015). In addition to ground motion analysis, crustal deformation analysis, which is useful for estimating tsunami sources, can also be performed using this type of finite element model (e.g., crust deformation analysis with 3-D island-scale high fidelity is realized); see Ichimura *et al.* (2016).

3.2 Structures

This section explains the automated model construction for structures which are used for engineering simulation. The methodology of automated model construction is important since various analysis models need to be made for structural seismic response analysis. As will be explained in detail, the methodology developed in this monograph is based on the present state of digital data available for the automated model construction. In general, design charts of the structures are not open to the public. This is a great disadvantage for the automated model construction. Digital data about the structural external configuration as well as digital data about the land use are available. The methodology is developed to easily use such digital data the quantity and quality of which are expected to increase.

3.2.1 Methodology of automated model construction

The basic methodology of automated model construction is *template fitting*, i.e., for a given template of an analysis model that has a set of model parameters of the configuration and material properties, a suitable value for each model parameter is determined using the available information; see O-Tani *et al.* (2014a and 2014b). As for IES, the use of digital data of city and urban areas in automatically determining the model parameters is the key process of template fitting because only limited data are available for the structures.

There are a variety of templates for analysis models of structural seismic response analysis; as explained in the preceding section, an analysis model ranges from a simple mass-spring system to a solid element model of high fidelity which can be analyzed by using FEM enhanced with HPC functionality. In general, as the complexity of the model template increases, we can make a more reliable estimate of the structural seismic response. However, we must determine more model parameters for a more complex model template since the number of model parameters increases as the complexity increases. There is a trade-off relation between the reliability and the complexity of the template of the analysis model. Hence, it is necessary to choose a suitable template among various templates for the automated model construction.

It is true that an analysis model that is constructed form a complex model template can make a more reliable estimate of the structural seismic response with higher spatial and temporal resolution. However, the reliability of the estimate depends on the correctness of the model parameters. An analysis model that uses wrong values for its model parameters produces an incorrect estimate of seismic response, even though it is constructed from a complex model template. It is not an easy task to correctly determine many model parameters using limited digital data. The quality of the analysis model is thus evaluated in terms of the following two elements: 1) the complexity of a template, from which the analysis model is constructed; and 2) the correctness of the model parameters, which are determined from the digital data.

For the given digital data of a structure, we can construct various analysis models by choosing a model template and determining the model parameters from the digital data. These models ought to share the basic dynamic properties, such as the natural frequency, the mode shapes, the damping factor, and the seismic capacity. This is the consistency of analysis models that are constructed for a common structure. For instance, the primary natural frequency can be computed for the simplest analysis model of a mass-spring model which uses only two model parameters, namely, the mass and the spring constant. Other more complicated analysis models ought to have a similar value of the primary natural frequency, provided that the primary natural frequency of the simple model is accurate. We can tune the model parameters of the analysis model made from a more complicated template, satisfying the consistency with other analysis model made from a simpler template.

The consistency of analysis models constructed for a common structure can be used to select a suitable model template among a variety of model templates. That is, starting from the simplest template, we continue to choose a more complicated template until the consistency with the previous analysis model is lost for the new analysis model constructed from the template. Due to the limitation of the available digital data, some model parameters cannot be determined or wrong values are chosen for determining some model parameters of the complicated template, which results in a loss of the consistency of the new analysis model.

It should be noted that the use of consistency in choosing a model template is based on the presumption that the analysis model that is constructed by choosing the simplest template and determining a few model parameters from the limited digital data have accurate dynamic properties. This presumption is considered to be true since it is surely easier to construct a reliable model of the mass-spring model, compared with other more complicated analysis models. Besides the presumption, we can take advantage of the engineering information about the dynamic properties to examine the consistency. For instance, the primary natural frequency is correlated to the height of a residential building. Using this correlation, which is regarded as engineering information, we can examine the correctness of the primary natural frequency of the analysis model that is constructed from the simplest template.

3.2.2 Procedures of automated model construction

According to the methodology explained in the previous subsection, we develop basic procedures of automated model construction using digital data of limited quality and quantity as follows: 1) preparing a few model templates of different complexity; 2) choosing a suitable model template considering the information obtained from the available digital data; 3) determining the model parameters if it can be reliably determined from the available digital data; and 4) for other model parameters, using the default value, which can be determined from the engineering information as well as the model parameters that are determined by the previous procedure; see O-Tani *et al.* (2014b). While these procedures seem reasonable, it is difficult to make a program that follows the procedures precisely. A program for automated model construction must be made for each type of structure, which has its own characteristics for the analysis models. Still, it is important to establish the basic procedures for automated model construction so that the modules developed for one type of structure can be reused for other types of structures.

As for the first procedure, the present system of IES does not prepare a large variety of templates for analysis models. This is because only limited kinds of digital data are available and only a few templates are used to construct an analysis model. In the near future, larger kinds of digital data are available for the structures, and we need to use a large variety of templates. Therefore, the program for automated model construction must be designed to have high extensibility so that it can handle the extension of templates.

The present system of IES makes the choice of a template in a deterministic manner (not in an evolving manner which starts from a simple template to a more complex template until the consistency of the constructed model is lost, as explained in the previous subsection) for the second basic procedure. This saves the efforts of developing a program but it is not certain that the chosen template is the most suitable for the given digital data. This is the limitation of the deterministic choice. As for the third and fourth procedures, the present system of IES makes the choice of determining the value of the model parameters from the available digital data or using the default value which is computed from the engineering information or other model parameters; the choice is made based on the availability of specific digital data. Like the second procedure, this deterministic choice saves programming efforts. Again, it is surely the limitation in constructing a reliable analysis model from a template. We need to develop a more sophisticated program for the second, third, and fourth procedures.

A sophisticated program must have the functionality of evaluating the quality and quantity of the available digital data in constructing an analysis model for a chosen template. There are numerous choices for selecting a suitable template together with determining the model parameters for the selected template. It is necessary to quantify the result of choice so that we can determine the most suitable template and the most suitable model parameters. As explained in the preceding subsection, we need to adapt an evolving algorithm[1] in which the simplest template is first used to construct an analysis model and a more complex template is used until the consistency of the constructed analysis model is lost.

3.2.3 Automated model construction of residential building

Data sources for automatically constructing analysis models of residential buildings are limited. The following two data sources are available in Japan: 1) 3D map, which is originally used for automobile navigation systems, is a useful data source, which stores relatively accurate data about the external configuration of each residential building; and 2) land registration, which stores data about the construction year, structure type, etc., for each residential building in Japan. The basic flow of automated model construction that uses these two data sources is presented in Fig. 3.3.

A difficulty in combining different data stored in these two data sources is that the data are scattered and cannot be readily used for constructing an analysis model. To be specific, the data stored in the 3D map is a set of polyhedral which are not connected to a residential building; visualizing all polyhedra individually makes a perfect three-dimensional view and there is no need to connect polyhedra to each residential building. By correctly connecting the polyhedra to a residential building, we can extract information about floor arrangement and structural members. Land registrations use a lot number and a house number to relate the stored data to a residential building. It is necessary to convert these numbers to more accurate geographical information such as latitude and longitude. The accuracy of the converted geographical information must be as high as that of the 3D map to combine the data of land registration to the 3D map. It is essentially important to maximize the amount of extracted data and to minimize the wrong combination of the two data, so as to combine the two data sources.

The 3D map is commercially available, and it is readily used as it is of the GIS format. The key feature of the 3D map is the high accuracy of the stored data about the external configuration of residential buildings so that the map can be used for an automotive navigation system. The data structure of the 3D map is simple, i.e., a huge set of polygonal faces that represent an external surface of a residential building. As mentioned, polygonal faces are not connected to the residential building which the polygonal faces represent.

[1]Implementing this algorithm is the most challenging task in developing a program for automated model construction.

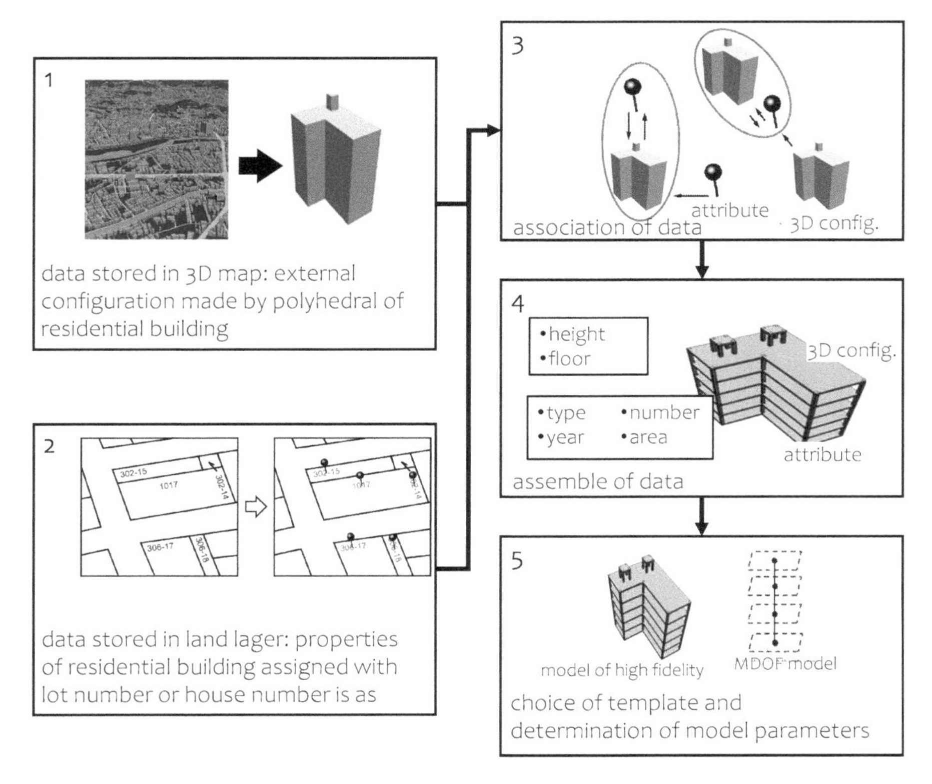

FIGURE 3.3: Basic flow of automated model construction for residential building.

For automated model construction, a robust method of choosing a set of polygons for one residential building from a set of independently recorded polygons has been developed. A method of computing the floor area and extracting information about the floor arrangement has been developed. Such floor-wise information is essential for constructing a variety of analysis models ranging from a multi-degree-of-freedom model to a solid element model of high fidelity. For instance, a multi-degree-of-freedom uses the floor number to determine the number of mass points, and the mass is determined by the floor area.

Polygons that are obtained by casting the extracted polyhedra of a residential building on a floor plane often have gaps and overlaps among them. A polishing process that uses techniques developed in computational geometry is needed to correct such gap and overlaps, and the floor arrangement is completed. Template fitting is used to this process. Thus, for a set of floor templates, we select the most suitable template and determine the model parameters of the template. We can prepare the simplest template of a rectangular floor, which has model parameters, and the complicated floor template, which includes parameterized walls and columns in it.

In Japan, the operation of land registration is based on Article 380, Paragraph 1, Local Tax Law, "Municipalities must have a land registration (or real property registration) to clarify the status of property and the price of property, which is used as the base of the property tax." It is a nationwide data source. According to Article 44, Paragraph 1, Real Estate Registration Law, it is stipulated that the registered items of a residential building are included as attribute information of the building; the registered items are the name of the city, ward, county, town and village, a lot number, a house number, the building type and structure, and the floor area.

It should be noted that there are cases wherein one residential building has multiple data in land registration. For instance, a condominium is stipulated as a compartmentalized ownership building according to the Law Concerning Divisional Ownership of Buildings. The ownership is established for each room; hence, one condominium has multiple data for each room in the land registration. It is thus necessary to combine multiple data, which are made owner-wise or room-wise to single structure-wise data.

The location of a building recorded in the land registrations is usually confirmed using the following two steps: 1) confirming the location of a residential building by referring to the official map specified in Article 14, Paragraph 1, Real Estate Registration Law or the map specified in Article 14, Paragraph 4, Real Estate Registration Law; and 2) confirming the arrangement of the residential building by referring to the building location map specified in Article 14, Paragraph 1, the Real Estate Registration Law. Manual works are usually required for these two steps, since the two maps are not available in the digital form. Instead of using the two maps, many municipalities have released a lot number reference map and have opened it to the public. While the lot number reference map is not as accurate as the two maps that are specified by Real Estate Registration Law, this map is digital. Hence, we can use the lot number reference map to identify a residential building while extracting its data, which are stored in the land registration.

For each residential building, the data about external configuration and the data about building attribute information are respectively extracted from the 3D map and the land registration, and they are combined as building information data. No duplication should be made in combining the two data, by relating them based on the most accurate geographical information. The geographical information for the land registration data is determined by first assigning the data to the lot number reference map and then computing the geographical information of a representative point of the lot on the map; the point is chosen as the center of the lot number shown on the map. For a residential building that is closest to the point, the land registration data are related to the 3D map data.

It is important to examine the correctness of the procedures mentioned above of relating the registration data to the 3D map data. There is a possibility of making an error in determining the geographical information of the land registration data or in calculating the distance of the land registration data to the 3D map data for a targeted residential building. At this moment, the reliability of automatically relating the data stored in the two data source is as high as the manual work. For more reliable automated relating, it is necessary to improve the accuracy of the lot number reference map or to digitize the two maps (the land registration map and the building location map) that are specified by Real Estate Registration Law.

3.2.4 Automated model construction of road bridge

It is possible to automatically construct a reliable analysis model for newer road bridges to which the digital data of the design charts are available; template fitting explained in the preceding subsections is applicable to such structures. A majority of bridges do not have such digital data; some have lost even the printed materials of the design charts. An analysis model ought to be constructed based on the presumption that sufficient digital data are not available. The presumption used here is that the model parameters can be determined using design regulations and guidelines since a bridge is constructed according to them.

In general, road bridge consists of the following three parts: 1) a superstructure that consists of a deck, girder, parapet, etc.; 2) a substructure that consists of a pier, column, abutment, etc.; and 3) a foundation that consists of footing, piles, etc. An analysis model for seismic response analysis must be made for each of the three parts; for instance, the simplest setting is that a mass and a spring are used for the superstructure and the substructure, respectively, with the foundation being ignored; the most sophisticated setting is that the three parts together with the surface ground are modeled by solid elements to construct an analysis model of high fidelity so that the seismic response is computed by fully considering the interaction effects of the three parts as well as soil-structure interaction. While there is no single common analysis model, a few representative templates of an analysis model are available for each part of the bridge.

Similar to the automated model construction of a residential building, a suitable template is first chosen by considering the complexity of the three bridge parts. In choosing a template, we convert all the available digital data to a common form, called *common modeling data* (CMD) for bridge information. As shown in Fig. 3.4, the CMD is assembled from the following three data sources: 1) bridge data from the design charts given as the digital documents or printed documents; 2) references such as design regulations, design database, bridge regulation, and road regulation; and 3) road data, which are included in land use databases of road information, region information, elevation data, geological data, and information about the river across which the bridge is built. The references are converted to derivatives such as empirical relations or presumptions which are used to determine[2] the model parameters. The road data are converted to derivatives of locations, span, elevation, or ground

[2]Some parameters of a template can be determined from the bridge data but there are other parameters that cannot be determined from the bridge data. The derivatives made from the references and the road data are thus used to guess the suitable values of these undetermined parameters.

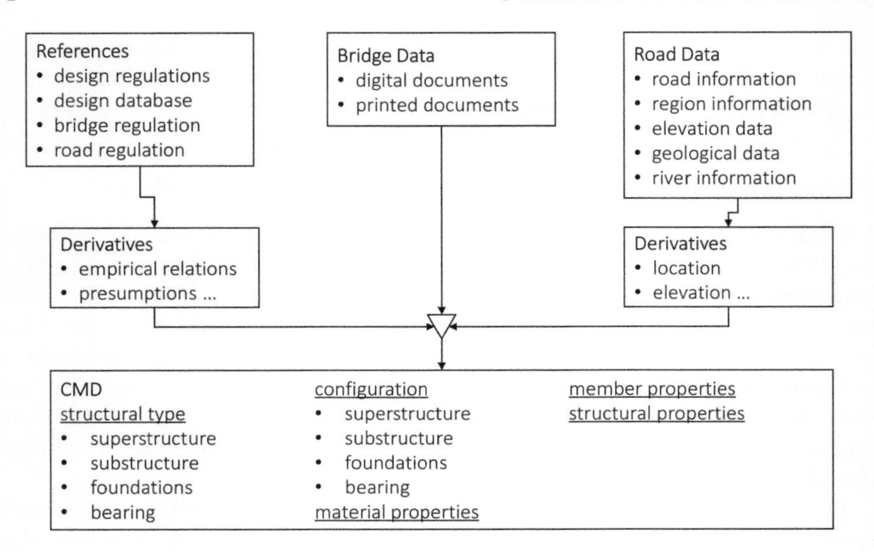

FIGURE 3.4: Overview of CMD.

conditions. CMD is first assembled from the bridge data, and information missing in this CMD are added using the two derivatives.

As shown in Fig. 3.5, CMD is used to process the digital data for automated model construction. The following two functionalities are used in data processing: 1) the functionality of choosing a suitable template based on the evaluation of the bridge information stored in CMD; and 2) determining the model parameters of the chosen template using the bridge information. The CMD is determined by assembling various data sources, and is used to determine a suitable model, from primitive to sophisticated, by first choosing a suitable template and then determining the model parameters. It should be mentioned that CMD is required to have another functionality of validating the constructed model. As for seismic response analysis, we can examine the dynamic properties of the constructed model, such as the natural frequency and mode, by comparing the empirical relations of these properties and presumptions.

Fragmented bridge information needs to be fulfilled using the derivatives of the empirical relations and presumptions which are based on the reference data such as design regulations, bridge regulation, road regulation, and design databases. A major data source is a digital road database which stores the basic data of almost all roads in Japan. Also, a land use database is used, which includes road information, region information, elevation data, geological data, and river information. The data stored in the land use database are converted to derivative information such as the location, span, elevation, and ground conditions of a bridge.

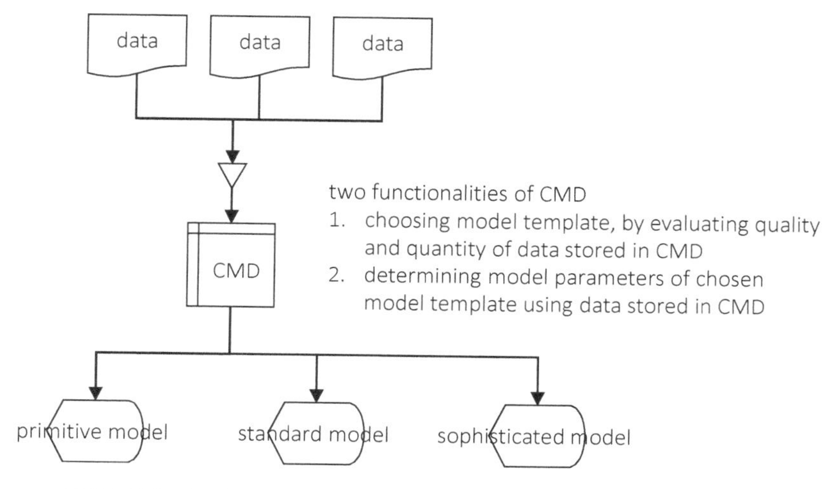

FIGURE 3.5: Process of determining model template using CMD.

The algorithm of constructing the CMD of a road bridge is explained[3] in Fig. 3.6; a) is for the algorithm for making a superstructure; b) is for the algorithm of substructure and foundation, which is much simpler than a). A gray box, a white box, and a diamond stand for data, process, and conditioning, respectively.

For a segment of the road that is categorized as a bridge, the length L and the curve Φ_{ave} are first determined from the data about the road node and link, as shown in the top of Fig. 3.6a). The span number N_{span} and the pier number N_{pier} are computed from L and Φ_{ave}. The bridge elevation at the both ends $z(1)$ and $z(N_{sub})$ is determined from the data about the road elevation, and the elevation difference from the ground surface, dz, is computed. Based on this bridge information, the superstructure type ID is determined, and the superstructure height h and the bearing conditions S are estimated; the processes of computing h and S are complicated as shown in the right side of Fig. 3.6a. The bridge width W_r is determined from these model parameters together with the data about the road node and link. The superstructure weight W_s is computed using W_r and h. Finally, the rigidity and capacity of the superstructure are estimated from W_s, as shown in the right bottom of Fig. 3.6a).

The data of the construction year are used to determine pier parameters related to its configuration, as shown in the left side of Fig. 3.6b. Shoe height and spring, h_{shoe} and K_s, are determined from the shoe condition estimated

[3]Stochastic modeling is omitted in the algorithm shown in Fig. 3.6 because it can be used in any part of the algorithm that determines a model parameter.

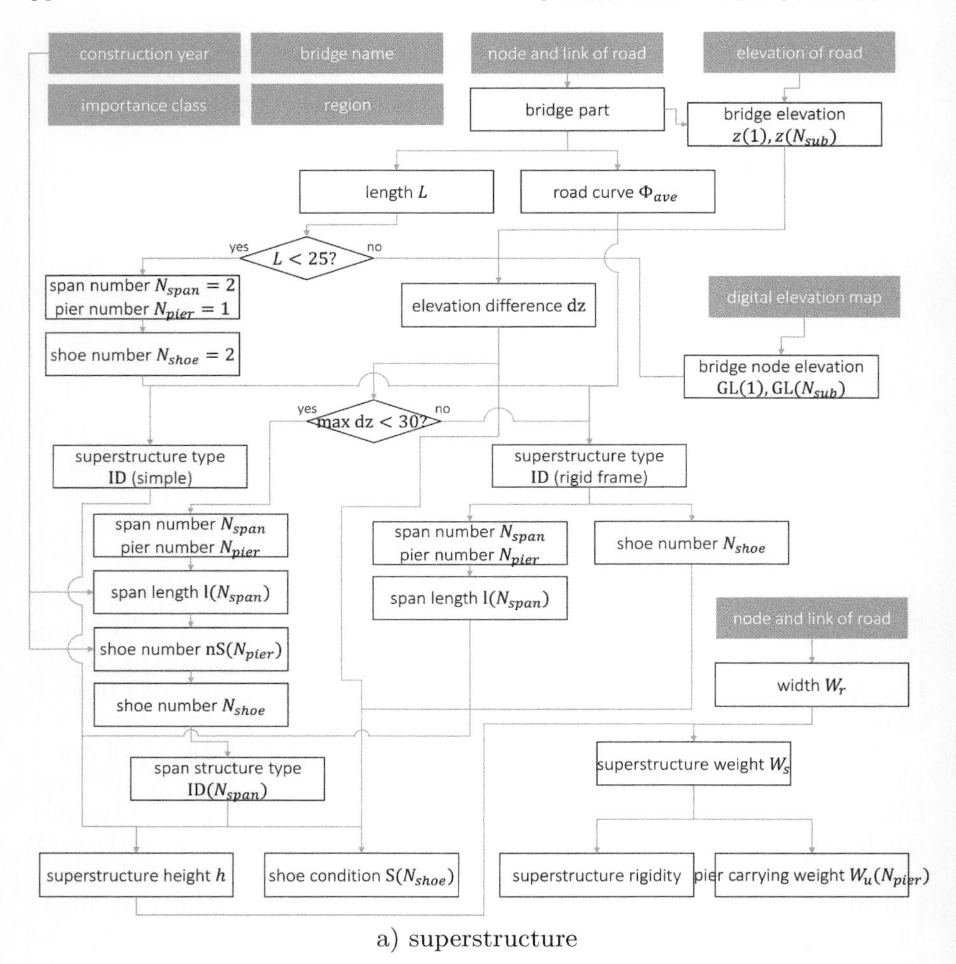

a) superstructure

in the superstructure algorithm. As for the ground, the data of AVS30 and the boring data are used to determine the ground type, from which the foundation type B_{type} is estimated. If the foundation type is pile foundation, foundation characteristics such as the foundation type, foundation length, and foundation spring are estimated. If the foundation type is not pile foundation, rigid connection between the foundation and pier is used, as shown in right bottom of Fig. 3.6b).

There are various possible procedures for converting a set of data stored in the CMD to the model parameters for a chosen template. The key issue in this conversion is to make the procedures most rational so that they are applicable to the case when the available data are limited. In the present system of IES, the rationality means the use of empirical knowledge. For instance, a standard design database that has stored bridge design charts for

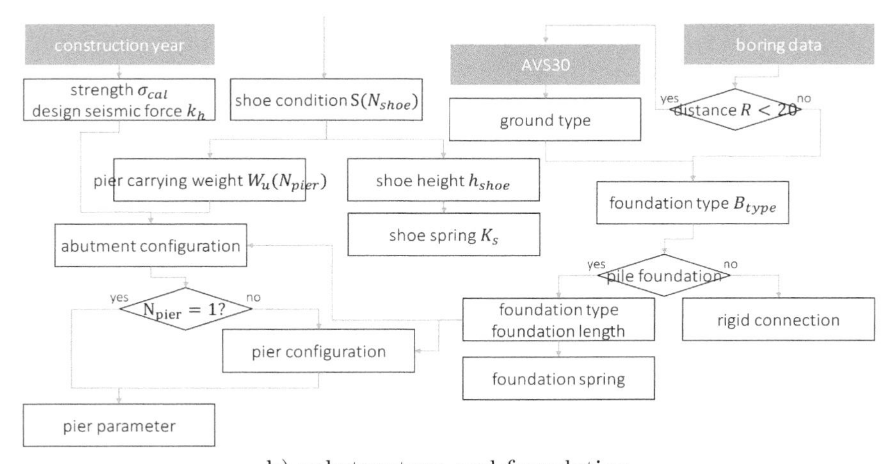

b) substructure and foundation

FIGURE 3.6: Algorithm of constructing CMD of road bridge.

several decades is used; for instance, a huge collection of actual design chart of the bridge cross-section is included in the standard design database, and we can use this collection to determine the model parameters of the cross-section when there is little data about it.

The standard design database of superstructures is used for constructing an analysis model. The database has 5 categories for superstructures and includes information about the road types (main major roads, major roads, auxiliary major roads, etc.), the area classifications, the cold districts and the bridge length classifications. There are 15 to 25 types of road width and cross-section configurations for each bridge type. In using the standard design database of superstructures, we must consider the number of available standard designs; for example, there are 2376 standard designs for the plate girder bridge, and it is impractical to develop a module for each of the standard designs. As a minimum effort, the present IES develops a module that outputs the cross-section configuration parameters from the input of the superstructure information that are related to the bridge height and the bridge span length, as well as the cross- section mechanical parameters (such as in- and out-of-plane rigidity, torsional rigidity and cross-sectional area).

The standard design database of substructures is mainly concerned with reinforced concrete piers of rectangular cross-section for land or oval cross-section for water. A large number of cross-section data are available for beams, columns, and footing, which are categorized on the basis of the superstructure type, the dead and live load classes, the beam width classes, the beam thickness classes, the ground types and the ground strength types. The empirical relations between the span length and the reaction forces and between the span length and the road width are included, and the use of these empirical

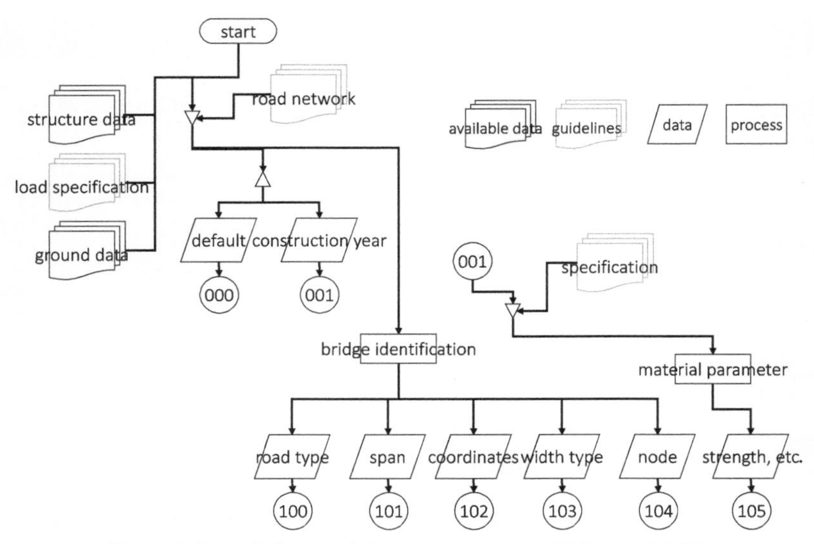

a) modules of determining parameters 000s and 100s

relations enables us to determine the substructure model parameters using the superstructure information mentioned above. Similar to the standard design database of superstructure, we must consider the number of available standard designs of the substructures; for example, there are 377 types of columns, beams, and footings of rectangular cross-section. Thus, it is impractical to determine the cross-section parameters for each type because there are many important cross-section parameters such as in- and out-of-plane rigidity, torsional rigidity, cross-section area, and moment and curvature for cracking, yielding, or failure. The present system of IES chooses 40 representative cross-sections of substructures and develops a module for each type which computes cross-section parameters using substructure information. As for the bearing, the present IES selects a fixed/movable/elastic bearing depending on the construction year if no data are available. The location of the bearing is rationally determined; for instance, a fixed bearing is used for the top of a shorter pier, when a pair of fixed and movable bearings is chosen.

In coding the modules of the automated model construction, we pay attention to the ease of improving and extending the module functionalities. For instance, the objects and functions used in the modules are named so that the meanings of the objects and functions are readily understandable; in particular, the input arguments and the output variables for an empirical relation must be named systematically to avoid making comments for explanation of the empirical relation.

Figure 3.7 presents a line-up of modules that are used to determine the model parameters of a few templates. The modules are allocated according to the flow of constructing a set of a superstructure analysis model, a

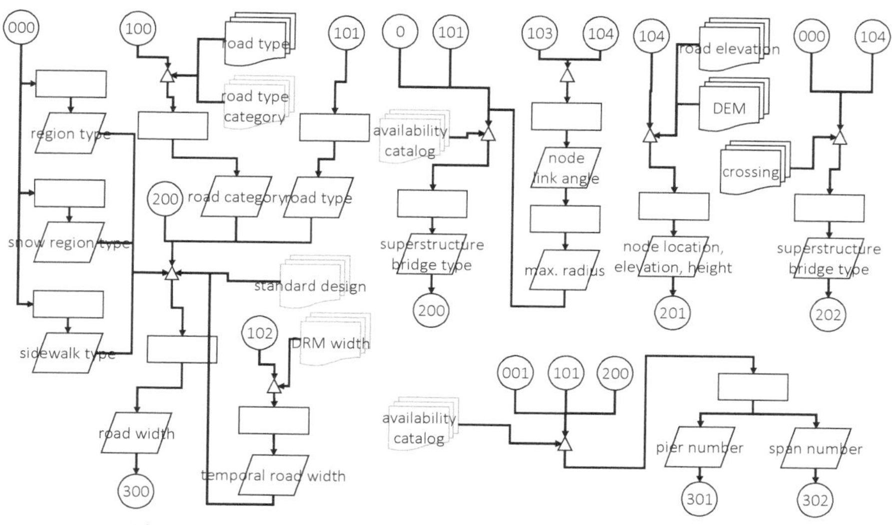

b) modules of determining parameters 200s and 300s

substructure analysis model and a foundation analysis model, and the parameters are classified as follows: 1) the 000s and 100s for the bridge basic properties; 2) the 200s and 300s for the superstructure configuration; 3) the 400s for the span; 4) the 500s for the substructure pier; 5) the 600s for the substructure load; 6) the 700s and 800s for the foundations; and 7) outputting files for analysis models, visualization. and parameters used. It is easily seen what model parameters are determined from the data using the process of converting data, and what parameters are speculated by combining the available data and references.

Figure 3.7a) shows two modules. The module on the left determines parameter 000 or 001 for a default construction year or the construction year, respectively. This module assembles bridge data, load specification, ground data, and data about the road network, and, judging the assembled data, decides to determine parameter 000 or 001; as will be shown, parameters 000 and 001 are key model parameters which are often used to determine other model parameters. This module also determines parameters 100 to 104, which correspond to the basic properties of a bridge, such as the road type, span, and coordinates, via a process of bridge identification which uses the assembled data. The module on the right determines parameter 105 for the material strength, first assembling parameter 001 of the construction year and specification data, and then applying a process for the material parameter identification.

There are five modules in Fig, 3.7b). The module shown on the top center determines parameters 200 for the superstructure bridge type using parameters 101, 103, and 104 when parameter 001 of the construction year is determined. The two modules on the top right determine parameters 200 and 201 for the superstructure bridge type and node information, respectively, when

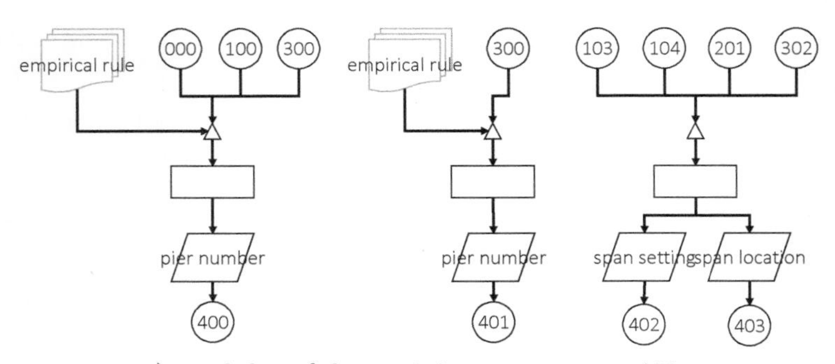

c) modules of determining parameters 400s

parameter 000 for a default construction year is used instead of parameter 001 for the construction year. The module on the left determines parameters 300 for the road width through complicated sequences of processes, using parameters 000, 100, 101, and 200, together with additional data of the road type, road type category, standard design, and Digital Road Map (DRM) width data. The module on the right bottom determines parameters 300 and 301 for the pier number and the span number, respectively, using parameters 001, 101, and 201 as well as additional catalog data.

Figure 3.7c) shows three modules for parameters 400s. The module on the left and center determines the pier number. Note that parameter 300 is temporally determined for the pier number and that the modules here finally determine the value. If parameters 000, 100, and 300 are determined, the module on the left is used to determine parameter 400 for the pier number by applying an empirical rule which needs the three parameters. If only parameter 300 is determined, the module in the center is used to determine parameter 401 for the pier number using another empirical rule which uses parameter 300 only. The module on the right determines parameters 402 and 403 for the span strength and the span location, respectively, using parameters 103, 104, 201, and 302.

Figure 3.7d) shows four modules for parameters 500s, which are slightly more complicated than those shown in Figure 3.7c). The two modules on the left are simple; the module on the left end determines parameter 500 for the maximum span distance using parameter 302 and 402, and the next module determines parameters 502 and 503 for the span setting and location, respectively, using parameters 001, 201, 301, 302, and 402. The two modules on the right are complicated; the module in the center determines parameter 504 for the pier height, using parameters 201 and 403 via a few processes; and the module on the right determines parameters 505 and 506 for the ground condition and the span location, respectively, using parameters 000 and 403 together with additional data of AVS30 and the bridge data. The last module is complicated since it takes into consideration the ground condition.

In Fig. 3.7e), three modules for parameters 600s are presented. The module on the left determines parameter 600 for the pier beam thickness using

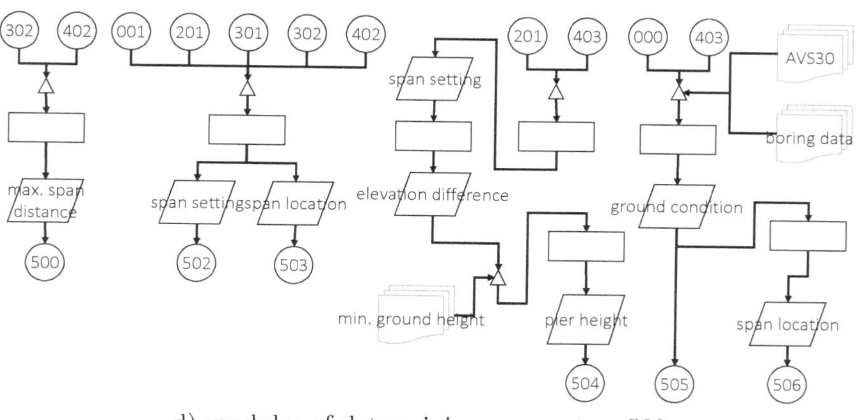

d) modules of determining parameters 500s

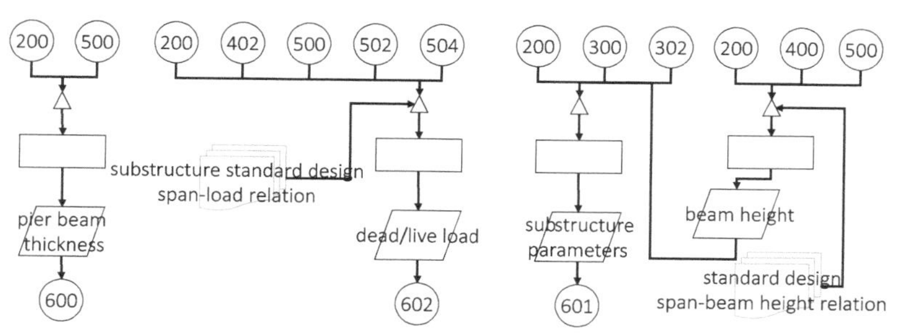

e) modules of determining parameters 600s

parameters 200 and 500. The module in the center determines parameter 602 for the dead/live load using parameters 200, 402, 500, 502, and 504, together with information of the substructure standard data and empirical relations between the span and the load. The module on the right determines parameter 601 of the substructure parameters mainly using parameters 200, 300, and 302. Parameters 200, 400, and 500, information about standard design, and empirical relations between the span and the load are used to determine the beam height, which is used as supplementary information to determine parameter 601.

Parameters 700 and 701 for the substructure parameters are determined by the two modules shown on the left of Fig. 3.7f). If parameters 200, 301, 400, 401, 504, 505, 600, and 602 are determined, the module on the left determines parameter 700, using additional information of the substructure standard design and assumed relation for the cross-section of the substructure. If these parameters are not determined, the module in the center determines parameter 701 using two parameters 502 and 602, together with the assumed relation for the cross-section of the substructure. The module on the right determines

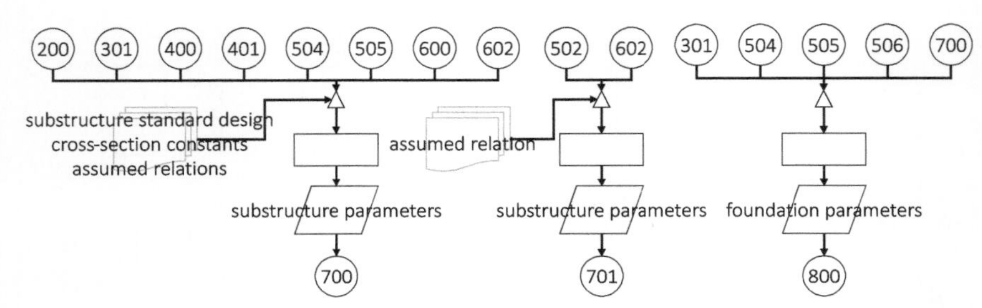

f) modules of determining parameters 700s and 800s

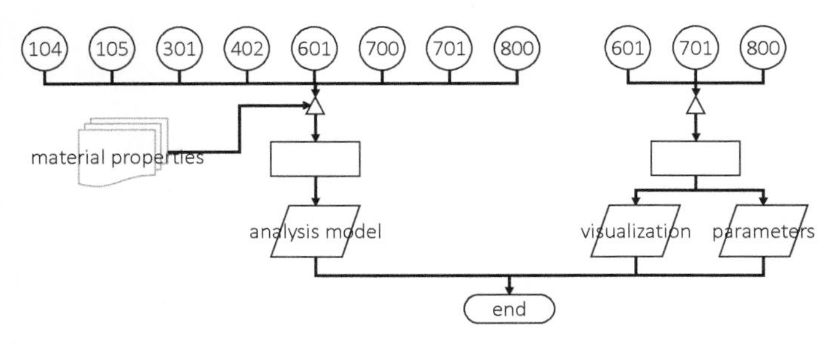

g) modules for outputs

FIGURE 3.7: Modules of determining model parameters of analysis model of road bridge.

parameter 800 for the foundation parameter using parameters 301, 504, 505, 506, and 700.

Figure 3.7g) shows the last module, which is located at the end of the line-up of modules; the line-up starts from Fig. 3.7a). An analysis model for the three parts of a bridge is determined using parameters 104, 105, 301, 402, 601, 700, 701, 800, and output files for visualization, and the model parameters are created by using parameters 601, 701, and 800.

3.3 Evacuation environment

The quality of the environment model significantly influences both the quality of the agent's behavior and the computational performance. A high-quality model of the environment significantly contributes to the smooth movements of the agents, significantly reduces the complexity of the constituent functions $g^{\cdots}$, and reduces memory requirements and execution time. As an example, Fig. 3.8a shows two models of the same environment generated using the vector data of road edges, and the vector data of the road centerlines and widths as the inputs. Although grids of comparable quality can be generated from both these sources, the road center line graph generated from the former

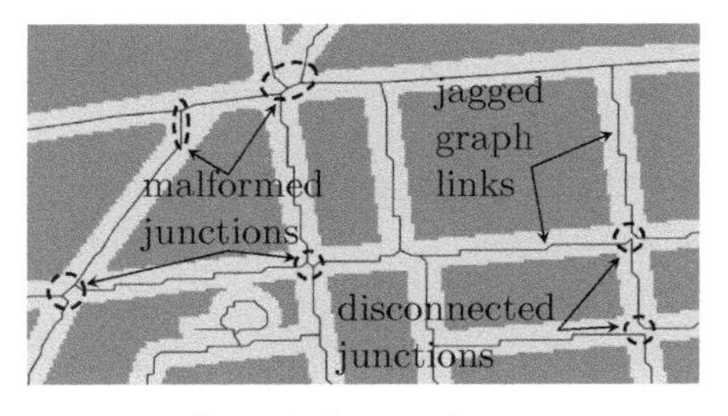

a) road edges as input

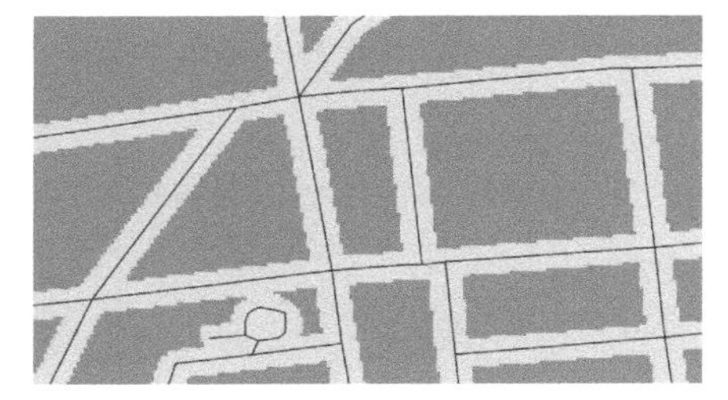

b) road center lines and widths as input

FIGURE 3.8: Comparison of grid and graph generated using road edges and road center lines as inputs.

data source is quite poor. Due to difficulties encountered in identifying the road center lines from the road edge data, the road center line in Fig. 3.8a is generated by thinning (Beeson *et al.* 2005) the roads in the raster image. The poor quality of the graph of Fig. 3.8a leads to several problems: requires 5 to 10 times more links and nodes, increasing the memory consumption and path planning time; requires complicated logic to make vehicles move along realistic trajectories at junctions, multiple lanes, and lane changing; requires extra logic to make pedestrians take sidewalks. A high-quality model of the environment eliminates these problems, increasing the productivity. This subsection briefly explains the main steps involved in the generation of the graph and the grid environment for the evacuation simulator.

3.3.1 Automated construction of grid and graph

The main steps involved in generating our environmental model can be summarized as follows:

1. Extract necessary GIS data.
   (a) poly-line data of road center lines including road widths and other available information like the number of lanes, flow direction, etc.
   (b) closed polygons forming the footprint of buildings including other available data like the number of occupants, usage of each building, etc.
   (c) closed polygons forming water bodies, contour lines, public parks, parking areas, etc.
2. Merge the overlapping nodes of the extracted road center poly-lines, and generate the bi-directed graph of the road center lines $\mathcal{G}$. Assign the road widths and other available data like the number of lanes to each link, and remove the relevant directed links corresponding to one-way roads. A kd-tree (Bentley *et al.*, 1975) of the poly-lines' nodes can significantly reduce the node merging time when processing networks with millions of nodes.

3. Create a raster canvas such that each pixel represents the desired spatial resolution (e.g., 1 m×1 m). Pixelate each graph link, setting the width of the pixelated line according to the road width, and insert into the canvas. A graphic library such as Cairo[4] or GIS software such as QGIS can be useful for generating a raster image.
4. Pixelate closed polygonal regions such as buildings and water bodies, and insert into the canvas. A computational geometry library such as CGAL[5] would be useful for calculating the geometric properties such as the area of these closed polygons.
5. When including earthquake damages, the foot prints of buildings are expanded according to the expected level of damage.
6. When including tsunami inundation, series of grids representing the state of the environment at suitable time intervals are generated, updating the cell states according to the progress of the inundation.
7. The number of maximum evacuees that can be accommodated in evacuation areas such as tall buildings and high grounds are included in the graph nodes located inside the corresponding evacuation areas.
8. Include other required information in the canvas using an image editor and construct the grid model G by converting the canvas to a 2D array.

We used the GIS data available from the Geospatial Information Authority of Japan as the data source of roads, buildings, water bodies, and contours, while the data on public parks and car parking were obtained from OpenStreetMap. Following the above briefly explained steps, our automated preprocessor could generate the grid and graph of a 32 km×18.4 km region (588 km^2) of central Tokyo, including each road, each building, water bodies, and major parks, in a few minutes.

3.3.2 Approximating vehicle trajectories at intersections

Junctions are a major source of delays and congestion even in everyday life. The chances are high for major earthquakes to render the traffic signals inoperable, causing significant traffic congestion. Further, since the pedestrians are given priority during mass evacuations, junctions can become severe bottlenecks, especially on narrow roads of congested cities. In order to capture such conditions reasonably well, we automatically generated realistic vehicle trajectories at junctions and used a realistic approximation for the speed profiles of cars along curved trajectories at the junctions; see Section 2.3.4.

Analyzing vehicle trajectory observations by Alhajyaseen *et al.* (2013) we found that vehicle trajectories at most common intersection geometries, except U-turns, can be approximated using B-spline (Piegl *et al.* 1997) with knot

[4]https://www.cairographics.org/
[5]https://www.cgal.org

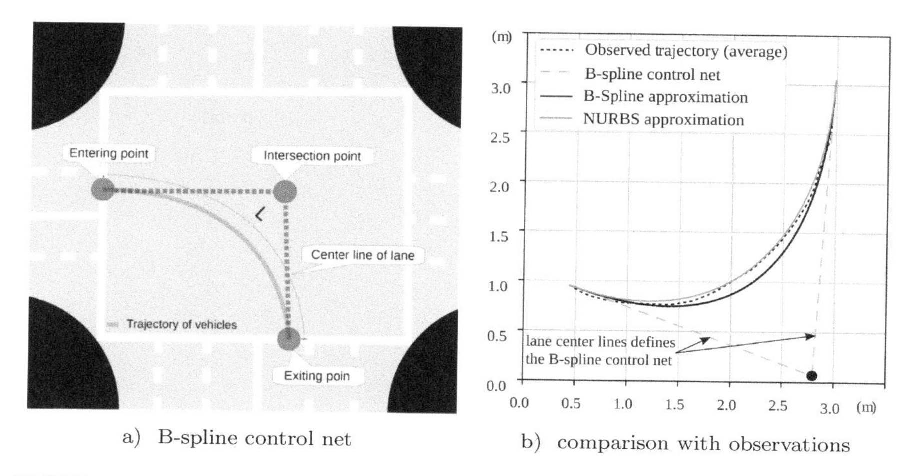

a) B-spline control net b) comparison with observations

FIGURE 3.9: Car trajectories at junctions are approximated with 3rd-order B-spline. B-spline's control net can easily be defined using the center lines of the lanes, as shown in (a). (b) compares the average of the trajectory observations by Alhajyaseen *et al.* 2013 and B-spline approximation.

vector $[0, 0, 0, 1, 1, 1]$ and three easy to find control points: 1) the lane center at the entry to the intersection, 2) the intersection point between the center lines of the incoming and outgoing lane, 3) lane center at the exiting point; see Fig. 3.9a). Figure 3.9b) compares our 3rd-order B-spline approximation and the mean of the vehicle trajectory observations conducted by Alhajyaseen *et al.* (2013) at the Suemori Dori 2 intersection in Nagoya, Japan. The B-spline approximation is only about 10 cm off the observations, which is negligible for this particular application. Though we found that a more accurate approximation can be made using NURBS, we use B-spline since it is relatively efficient to evaluate the 3rd-order Bezier basis functions and easy to define the control net; see Fig. 3.9b.

Figure 3.10 shows the main steps involved in the generation of trajectories, using a cross-intersection as an example. To reduce the amount of computations involved in making car agents move along the B-spline trajectories, we made a piece-wise linear approximation of the B-splines curves using a few number of points on the B-spline curves and assuming the trajectory between the consecutive points to be linear; see Fig. 3.10d). The potential collision points of the cars on different trajectories are defined based on the intersection points of the B-spline trajectories; see Fig. 3.10e). Since computing exact intersection points of the B-spline curves is complicated, we use the intersection points of the above piece-wise linear approximations of the B-splines. We found that 12 points (i.e., 11 piece-wise linear segments) provide sufficiently accurate approximation for the trajectories.

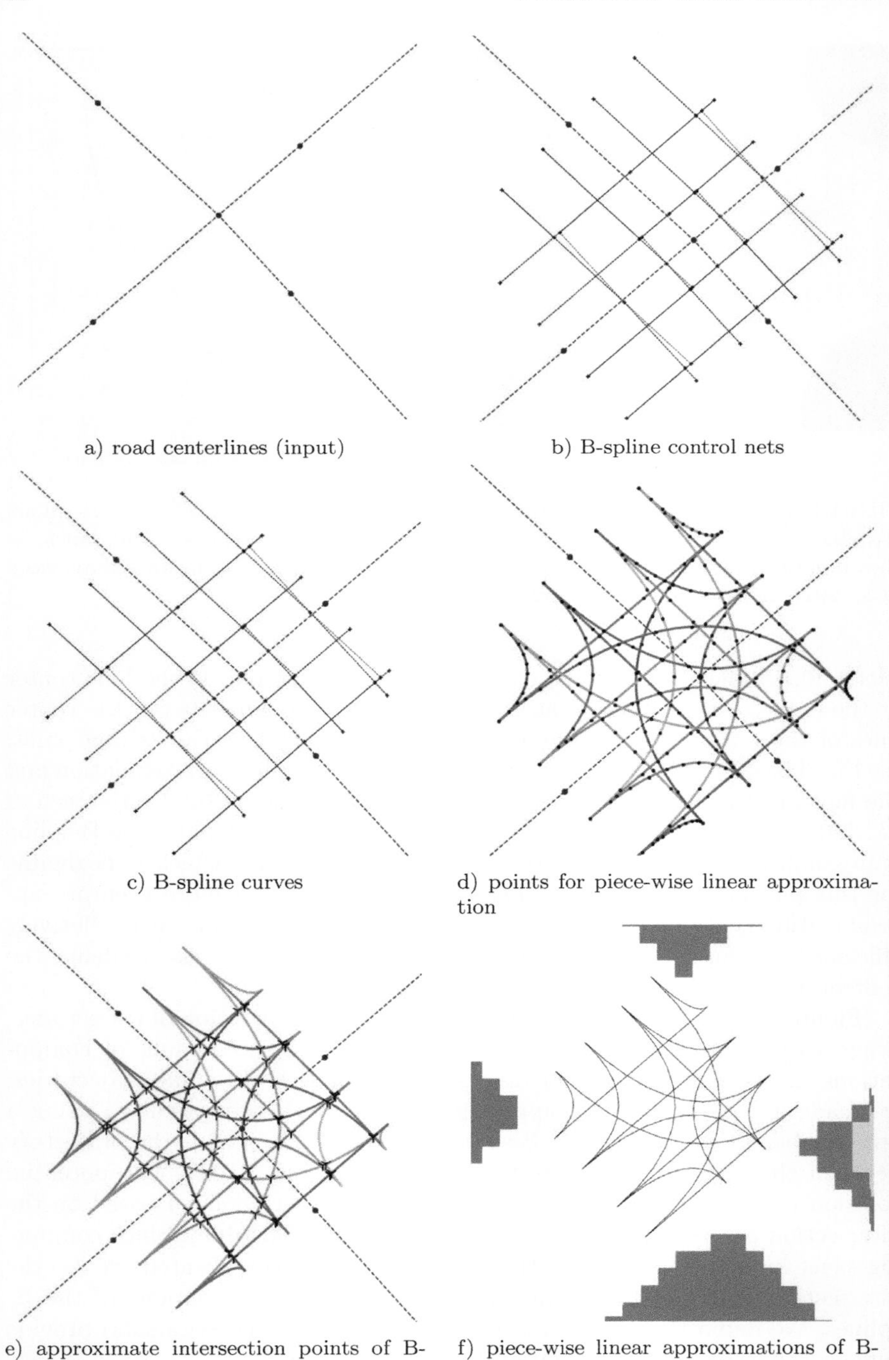

a) road centerlines (input)

b) B-spline control nets

c) B-spline curves

d) points for piece-wise linear approximation

e) approximate intersection points of B-splines

f) piece-wise linear approximations of B-splines shown on the grid (output)

FIGURE 3.10: Main steps of approximating car trajectories at a cross-intersection of three 2-lane roads and a single lane road.

4

Examples of Integrated Earthquake Simulation

CONTENTS

DOI: 10.1201/9781003149798-4

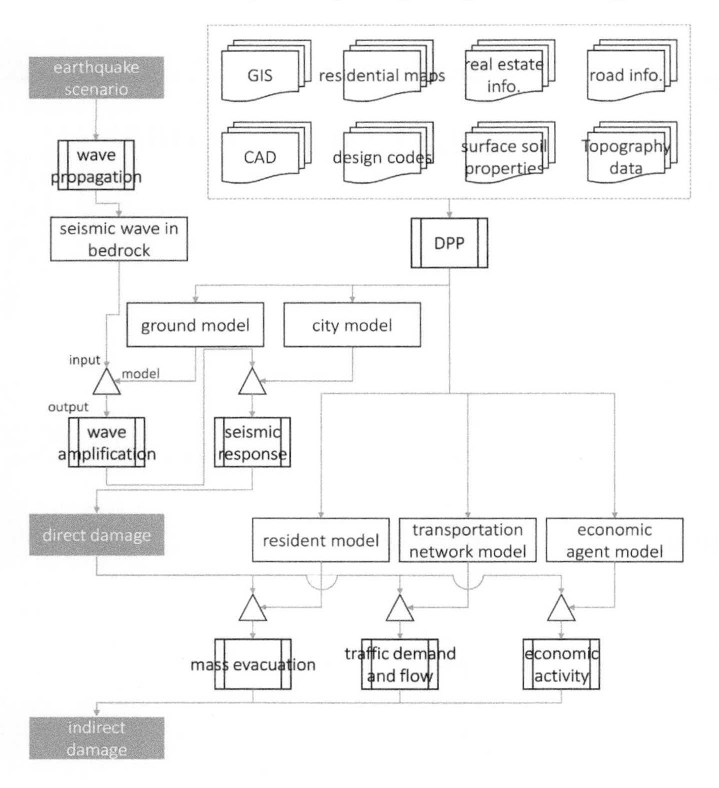

FIGURE 4.1: Basic flow of IES.

In this section, we present three examples of IES, which make a trial estimate of earthquake hazard and disaster using physics-based simulation, and the social simulation. While the first example studies large-scale physics-based simulation of Earth science simulation and earthquake engineering simulation, the latter two examples carry out the integrated simulation of Earth science simulation, earthquake engineering, and the social simulation for presumed earthquakes, namely, the Tokyo Metropolis Earthquake and the Nankai Trough Earthquake which, hit Tokyo Metropolis and Osaka area, respectively. The post-disaster response[1] obtained by social simulation is focused. The estimate of earthquake-induced damages in a wide area with high resolution enables us to carry out social simulation. The basic flow of IES used in the examples is presented in Fig. 4.1. The system for Integrated Simulation of Earthquake and Tsunami Hazard and Disaster 2020 provides an archive of research achievements related to the three examples.

[1]The estimate of the post-disaster responses is important to ensure better preparedness as well as to strengthen the social resilience for earthquake hazard. This is a major objective of developing IES; see Hori (2018).

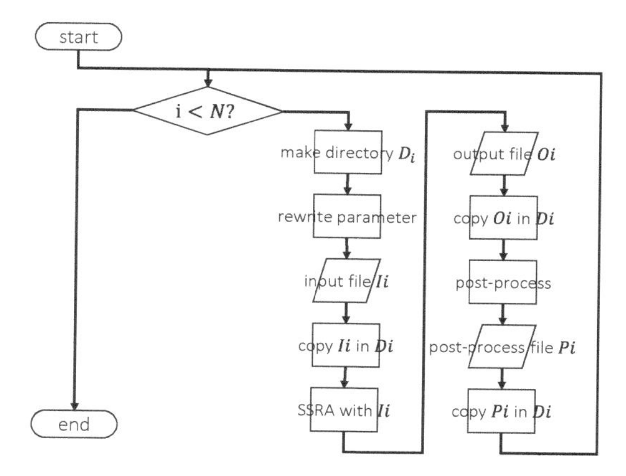

FIGURE 4.2: Object-wise multi-processing used in IES.

It should be mentioned that there are two classes of numerical analysis methods implemented into the current system of IES, namely, a parallel numerical analysis method enhanced with HPC capability and a serial numerical analysis method which is applied to numerous analysis models. The development of the first class is a challenging problem in the field of computational science. The realization of numerous simulations using the second class is useful for solving a large number of small problems. Figure 4.2 presents the flow of the second class adapted in the present system of IES; see Lu *et al.* (2014). While this flow is straightforward, it has a serious technical difficulty of parallel I/O related to the handling of a large number of input and output files; see Hori (2018).

4.1 Simulation of city blocks

4.1.1 Problem setting

The methods described in the previous sections enable us to automatically perform IES simulations using the city data. In this section, we analyze the seismic response of the city blocks in central Tokyo against the seismic design waves. IES simulations are specifically performed for city blocks A and B, which consist of ground amplification simulations using three-dimensional nonlinear ground amplification analysis and building simulations using nonlinear time history structural response analysis. City blocks A and B consist of an office area, commercial facilities, and a dense residential area; therefore, it is important to estimate the possible damage in this area during an earthquake. City block A is 2000 × 2000 m in size with 13,275 buildings, while city block B is 3500 × 3500 m in size with 41,675 buildings.

Even for such a large city block, it is essential to ensure the quality of the simulation results. For example, if the information on urban structures is not complete, it is necessary to evaluate them considering the ambiguity. Here, we show an example of a Monte Carlo simulation of the entire city block A to account for the ambiguity of the structural parameters of all the buildings in the block. It is also important to consider the quality assurance of the simulation to ensure, for example, that the numerical solution converges without depending on discretization in a large-scale three-dimensional nonlinear ground amplification analysis. Here, we confirmed that the results of the ground amplification analysis converged sufficiently without depending on discretization in city block B. Since this IES simulation is an extension of (Ichimura *et al.* 2014) with additional simulation, see (Ichimura *et al.* 2014) for details of the problem setup and the implementation of high-performance computing.

4.1.2 Models

In this IES simulation, three-dimensional ground structure data and building shape data are used. For example, as shown in Fig. 4.3 for city block A with 2000 × 2000 m area, a 5-meter numerical elevation map (Geospatial Information Authority of Japan 2019) and an electronic geotechnical map (National Research Institute for Earth Science and Disaster Prevention (Geo-Station) 2021) of this area are used. This three-dimensional ground structure data consists of three layers, namely, sedimentary layer 1, sedimentary layer 2, and engineering basin. The material properties of each layer of the ground are determined by referring to the material properties of the earthquake ground motion observation points near the corresponding points (National Research Institute for Earth Science and Disaster Prevention (Strong-motion seismograph networks) 2021). In addition, the shape data of 13,275 buildings are given. In the same way, the three-dimensional ground structure data of the three layers and the building shape data of 41,675 buildings are given for city block B.

Applying IES on this city data, we first perform a three-dimensional nonlinear ground amplification analysis and then perform a nonlinear time history structural response analysis for each building using seismic motions calculated at the relevant point of the building to simulate the entire city block. The modeling and analysis methods are based on the methods described in previous sections. In particular, the ground amplification analysis is described here in detail. Since the three-dimensional soil structure has a large influence on the distribution of earthquake ground motions, ground amplification analysis is performed by nonlinear dynamic analysis using a three-dimensional soil structure finite element model. A three-dimensional finite element method using unstructured tetrahedral quadratic elements is used because it analytically

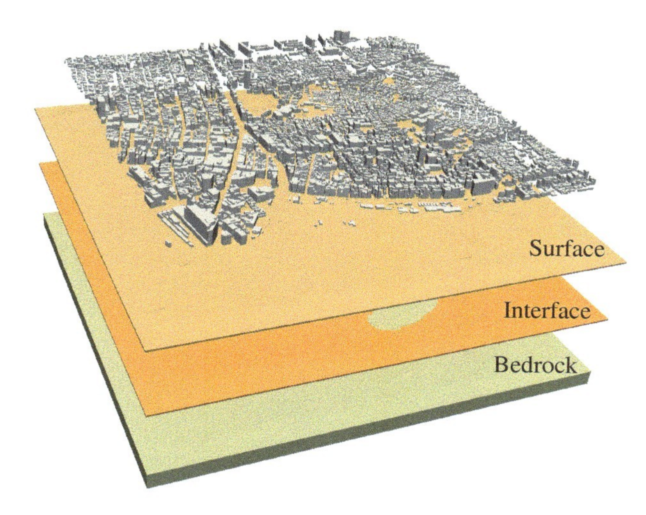

FIGURE 4.3: Ground and structure model of City block A.

satisfies the stress-free boundary conditions at the ground surface, accurately models the geometry of the ground, and is superior in handling local nonlinearities. In order to guarantee the accuracy of the numerical solutions, it is important to perform a finer discretization so that the numerical solution converges sufficiently without depending on the discretization. Here, a three-dimensional finite element ground model was generated using 0.66m unstructured tetrahedral quadratic elements to ensure that the numerical solution (acceleration waveform) converges sufficiently even when nonlinearization of the ground occurs.

As a result, the degrees of freedom of the generated finite element model become huge-10,755,536,091 degrees of freedom for city block A and 40,152,523,902 degrees of freedom for city block B. The modified Ramberg-Osgood model (Idriss Singh and Dobry 1978) and the Masing rule (Masing 1926) are used as the constitutive laws of the ground. From the bottom surface of this ground model, we input the earthquake ground motion observed during the Hyogo-ken Nanbu Earthquake (Japan Meteorological Agency 1995), which is also used as one of the seismic design waves, and perform seismic response analysis for city blocks A and B.

4.1.3 Simulation results

Figure 4.4 shows the seismic response of city blocks A and B using IES with three-dimensional nonlinear ground amplification analysis and nonlinear seismic structural response analysis of each building. In the 2000 $\times$ 2000 m city block A, 3D ground amplification analysis is performed with 10,755,536,091

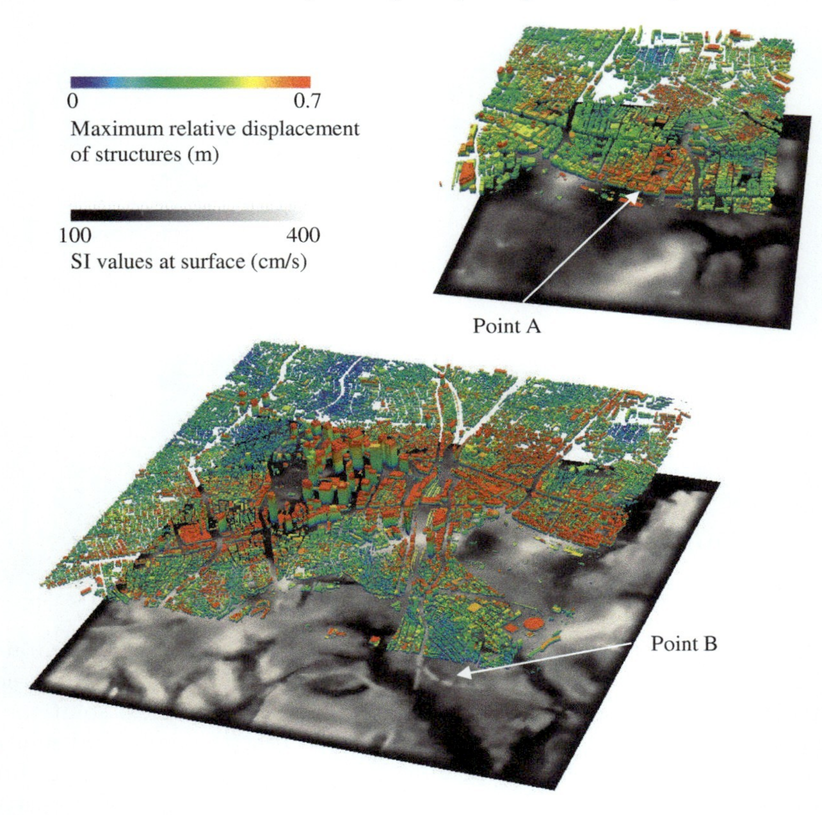

FIGURE 4.4: The results of the seismic response analysis (upper city block A, lower city block B).

degrees of freedom, and the calculated seismic motions are used for nonlinear time history structural analysis of 13,275 buildings. In the 3500 × 3500 m city block B, a three-dimensional ground amplification analysis with 40,152,523,902 degrees of freedom is performed, and the calculated seismic motions are used to perform a nonlinear time history structural analysis of 41,675 buildings. The black-and-white legend shows the distribution of the maximum seismic motion at the ground surface, and the colored legend shows the maximum response value of the building. The distribution of seismic motion is highly biased due to the three-dimensional ground structure. It can be seen that the response of the city is very complicated due to the fact that seismic motions differ from place to place and the characteristics of each building are different. It can be seen that they show completely different responses to the same input seismic motion.

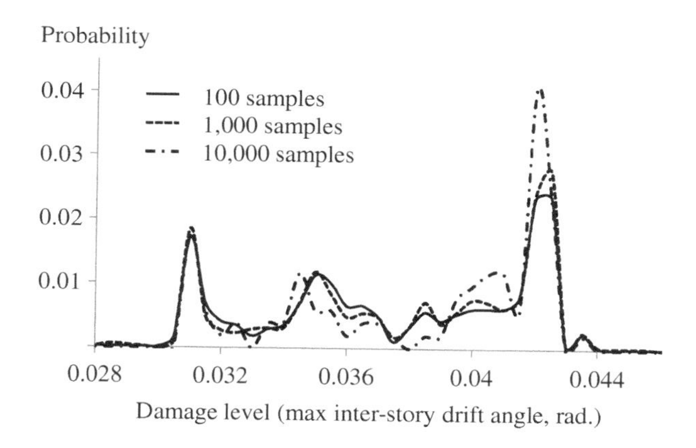

FIGURE 4.5: Probability density of response of two-story RC structure.

Due to the influence of the three-dimensional ground structure, the seismic motion at the ground surface differs greatly from place to place. In addition to the fact that the calculated ground motions at the building locations differ from site to site, the characteristics of each building are different, resulting in a very complex response for the entire city. This shows the effectiveness of IES, which is a bottom-up physics-based simulation process wherein simulation and data are coupled.

There are many structures in a city; thus, it is difficult to fully collect their structural parameters. Therefore, it is desirable to ensure the reliability of seismic response analysis of the structures by taking this ambiguity into account. Here, as an example, we consider that the material properties of the structures (stiffness of concrete) have ambiguity following a normal distribution, and show an example of a Monte Carlo simulation for all 13,275 buildings in city block A. In the Monte Carlo simulation, it is necessary to analyze the seismic response of each structure many times until the stochastic structural response converges.

As an example, Fig. 4.5 shows the results of a Monte Carlo simulation of a two-story RC structure located at point A in Fig. 4.4. The horizontal axis indicates the maximum inter-story drift angle, and the vertical axis indicates the probability of occurrence. 100 trials do not converge the stochastic response; thus, to obtain converged stochastic response, the number of trials is increased to 10,000. In this application example, Monte Carlo simulations were performed 10,000 times for all 13,275 buildings in city block A, which means that a huge number of structural response analyses, i.e., 13,275 $\times$ 10,000, were performed. However, using the 80,000 CPU cores (10,000 computer nodes $\times$ 8 CPU cores) of the K-computer (Miyazaki *et al.* 2012), we were able to perform this enormous analysis in a short time of 3 hours 56 minutes. This shows that IES applied by high-performance computing is also effective in improving the reliability of urban seismic analysis so as to account for ambiguities.

In numerical analysis, using a numerical simulation model with coarse discretization may result in inaccurate solutions. Therefore, in order to guarantee the accuracy of the numerical solutions, it is important to verify the quality of the numerical solution by conducting a numerical simulation using a model with a finer discretization so that the numerical solution converges sufficiently to the extent that it does not depend on the discretization. Here, we confirm the convergence of the numerical solutions in three-dimensional nonlinear ground amplification analysis for city block B shown in Fig. 4.4. In this analysis, we use a three-dimensional finite element model with 9,981,684,232 unstructured tetrahedral quadratic elements and a minimum size of 0.66 m with 40,152,523,902 degrees of freedom (10 billion element model).

We compared the results of the ground amplification analysis with those of larger element sizes. Specifically, the convergence of the solution is checked by comparing the results obtained by increasing the spatial resolution of finite element discretization of the ground finite element model by a factor of 1.5 (3 billion element model) and 3 (0.4 billion element model) with the results obtained by analyzing with 10 billion element model.

First, the distribution of the SI values (in cm/s) due to seismic motion on the ground surface of city block B is shown in Fig. 4.6. 10 billion element model results show that the distribution of SI values is complicated, reflecting the three-dimensional ground structure. From the difference between the 10 billion element model and the 0.4 billion element model, it can be seen that the 0.4 billion element model has large differences in many parts. On the other hand, from the difference between the 10 billion element model and the 3.0 billion element model, the differences become smaller, with a maximum difference of about 1%, indicating that a sufficiently converged solution has been obtained. In order to examine the convergence in more detail, the acceleration waveform at point B shown in Fig. 4.4 is checked. The results for the 10 billion element model, 3.0 billion element model, and 0.4 billion element model are shown in Fig. 4.7.

For the acceleration waveform, the numerical solution converges as the spatial resolution of the finite element model increases, and convergence is ensured at about 1% of the amplitude. As described above, the numerical solution using a coarse discretization model may not be sufficiently accurate. Therefore, in order to ensure the quality of the numerical simulation, it is necessary to conduct a simulation using a finer model to check whether or not the solution does depends on the discretization. Here, a large-scale nonlinear three-dimensional ground amplification analysis with 9,981,684,232 unstructured tetrahedral quadratic elements and 40,152,523,902 degrees of freedom is performed, which is also indispensable for the quality assurance of the simulation.

It was previously considered to be difficult to conduct simulations on such a large scale; however, by utilizing the analysis and model generation methods described in this book, such analyses have become possible. Note

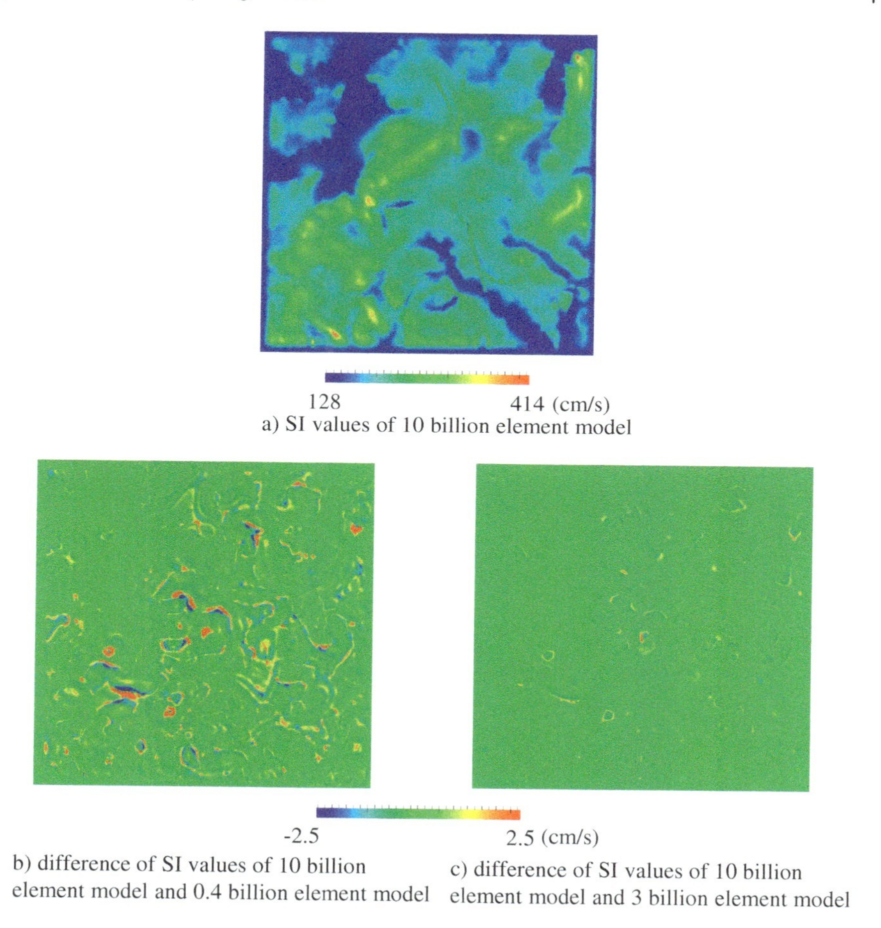

FIGURE 4.6: Ground motion distributions at surface.

that a large-scale nonlinear three-dimensional ground amplification analysis of 40,152,523,902 degrees of freedom with 9,981,684,232 unstructured tetrahedral quadratic elements was performed for 15,000 time steps with a time step increment of 0.001 seconds in city block B (at each time step, the equations are solved with a relative error of 10^{-8}). This analysis was performed in as fast as 6 hours 26 minutes on 663,552 CPU cores (82,944 computer nodes $\times 8$ CPU cores) of the full K-computer system (only 1.544 seconds per time step is needed on average). This shows that IES applied by high-performance computing is also effective in improving the reliability of the numerical solutions for urban seismic analysis, taking the convergence of the numerical solutions into account.

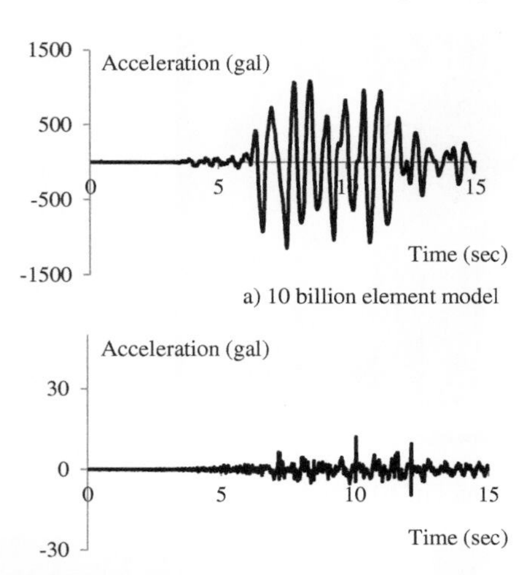

b) difference of 10 billion element model and 3 billion element model

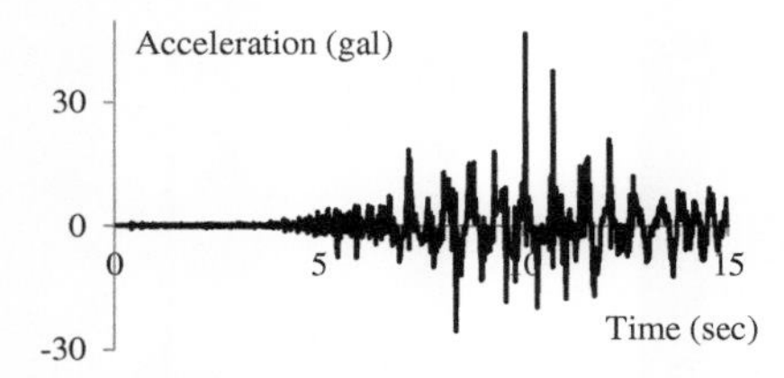

c) difference of 10 billion element model and 0.4 billion element model

FIGURE 4.7: Ground motion waveforms at surface.

In this section, we have shown the simulation of a city block using IES. The power of IES can be seen from the fact that it is able to analyze the complex response of a city by automatically generating a model using the complex city data and simulating a group of structures. On the other hand, one of the strengths of the IES applied by high-performance computing is that it makes it possible to guarantee the quality of the simulations, including those that were previously considered difficult due to the huge cost of analysis. In particular, as more and more data become available, and larger and higher quality simulations are incorporated into IES, high-performance computing, which is one of the points of discussion in this book, is a key to quality-assured urban earthquake simulation.

TABLE 4.1: Problem setting of IES-TME.

Earth Science	Earthquake scenario	1 scenario given by Cabinet Office Prediction
Earthquake engineering	Surface ground	3 layers for 10 x 10 km region, non-linear
	Structures	243,132 multi-degree-of-freedom models
Social science	Traffic flow	347,691 links
		558,572 packets of 15 vehicles

4.2 Tokyo Metropolis Earthquake

In this section, we simulate the Tokyo Metropolis Earthquake, which is considered to hit some parts of the Tokyo Metropolis within the next few decades. Several estimates[2] have been made for the hazard and disaster of this earthquake by applying empirical methods. For a more reliable estimate, we aim to study the applicability of IES to this big city and discuss the results obtained by the simulation.

4.2.1 Problem setting

To conduct IES for the Tokyo Metropolis Earthquake (IES-TME), we consider the following problem settings: 1) an earthquake scenario for Earth Science simulation; 2) analysis models of the surface ground and structures; and 3) analysis models of the road network and vehicles for social simulation. As explained in the preceding section, an earthquake scenario is a failure process on a fault plane, based on which the generation and propagation processes of earthquake waves are computed. The analysis model of the surface ground is used to compute the amplification process of ground motion, and the amplified ground motion at the site of a structure is used as the input to an analysis model of the structure. Traffic flow simulation in social simulation is performed in this example. These problem settings are summarized in Table 4.1.

As for an earthquake scenario, we adapt a scenario provided by the Cabinet Office, Japan; see Cabinet Office, Japan (2021a). There are a few epicenters expected for this earthquake, and we choose the one that hits the center of the Tokyo Metropolis. It should be noted that this earthquake is an intra-plate earthquake and can generate ground motions which include large portions of high frequency components. Strong ground motions are concentrated near the epicenter, and the high frequency components are harmful to many structures with a natural frequency of a few Hz. According to the conventional methods

[2]The Government of Japan and the Tokyo Metropolitan Government have made their own estimates of this earthquake. Their quotes are updated at intervals of less than ten years.

of evaluating earthquake hazard and disaster that use empirical relations, this scenario can cause considerable damage to the Tokyo Metropolis.

In the present example, the central part of Tokyo (which is surrounded by Yamanote Line) is chosen as the targeted area of the ground motion amplification, because this area is an alluvial plain which comprises a few large rivers; the underground structures are complicated and stronger ground motions are locally concentrated due to their topographical effects. As for the structural seismic response analysis, residential buildings are chosen as a target; the residential buildings include high rises in business districts, shopping and entertainment complexes, together with one or two-story houses.

Among the three numerical analyses of the social simulation that is implemented into the current system of IES, traffic flow simulation is carried out in the current example. Tokyo has a dense road network, which consists of various kinds of roads ranging from freeways of high quality to one-way narrow and curved roads. Numerous structures are located next to the roads, and the damage of these structures induced by ground motion can result in the loss of road capacity as they produce debris. For the accurate estimate of a possible congestion of the road network due to an earthquake, we must analyze all these processes[3] starting from the structural damage to the traffic flow in the road network with reduced capacity. As explained in Chapter 1, IES seeks to realize this analysis by combining earthquake engineering simulation with traffic simulation.

4.2.2 Constructed analysis models

An analysis model of the crust structure of the Tokyo Metropolis is constructed for FEM enhanced with HPC capability. The size of the model is $100 \times 100 \times 50$ km, and the DOF of this model is a few 10 million, which enables an earthquake wave propagation analysis of high spatial and temporal resolution; since the wave velocity of the crust is of the order of 1,000 m/s, elements of a few hundred-meter size are used to compute the frequency of up to 1 Hz. It should be pointed out that the resolution of ordinary earthquake wave propagation analysis is an order of magnitude larger than the present simulation.

As for the ground motion amplification process, a massive analysis model for FEM enhanced with HP capability is constructed. The most accurate analysis is needed to evaluate the topographical effects induced by complicated surface ground layers; hence, the model must have high fidelity. Due to the lack

[3]Conventional methods do not fully consider these processes of ground motions, structural damages, and traffic congestion. They can provide an approximate estimate of the traffic congestion based on the statistical data of the past earthquake traffic congestions.

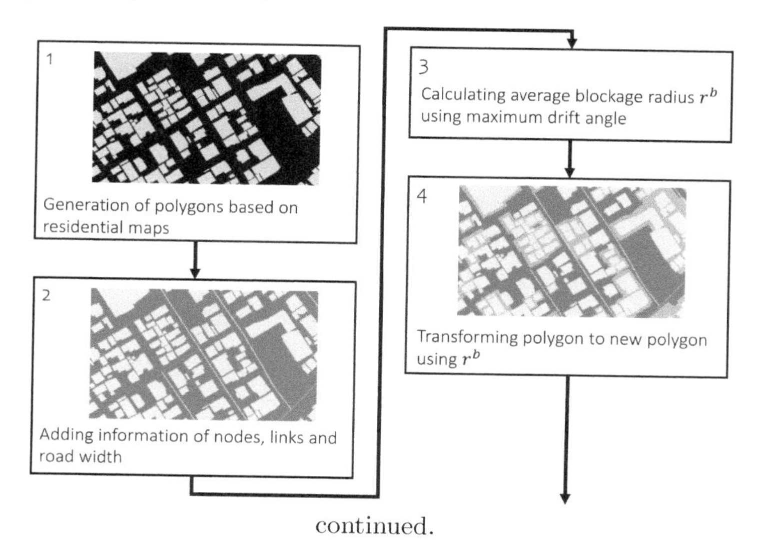

continued.

of a set[4] of boring data, however, we ought to compromise the construction of a high-fidelity model. Categorizing the soil layers into three, we construct the analysis model of the three layers, namely, a soft top layer, a middle hard layer, and a bedrock. The size of the model is $10,000 \times 10,000 \times 100$ m, and the DOF reaches 10 billion. To analyze the ground motion up to 5 Hz, a nonlinear constitutive relation of the Ramberg-Osgood model (RO model) is implemented into FEM. As for the structural seismic response analysis, the number of residential buildings in this area is 243,132, and a multi-degree-of-freedom model is constructed for each of them; the story number is used to construct the number of the mass, and the stiffness is determined according to the empirical relation between the first few natural frequencies and the building height. Linear springs are used for the model, even though these multi-degree-of-freedom models are readily converted into a nonlinear model.

As for the traffic flow simulation, we analyze the Central Metropolis Region, which covers the Tokyo Metropolis with the surrounding Kanagawa, Saitama, and Chiba prefectures. This region has a dense road network. An analysis model of 144,805 traffic nodes and 334,026 traffic links are made for the network. The number of vehicles using this road network is 26,646,532 per day, and traffic flow of 6 hours is analyzed in the traffic simulation.

To connect the earthquake engineering simulation of the structural seismic response analysis to the social simulation of traffic flow, we must change the road network model according to the structural damage. When structures are damaged, the nearby road segment loses its traffic capacity. As shown in Fig. 4.8, the procedures take the following seven steps: 1) generating a polygon

[4]Tokyo has a large set of boring data, which have been accumulated for a few decades. However, the density of the boring data is not sufficient to make an analysis model of high fidelity for the surface ground layers; for instance, there are many cases which display some inconsistency in the order of the soil layers between the adjacent boring data.

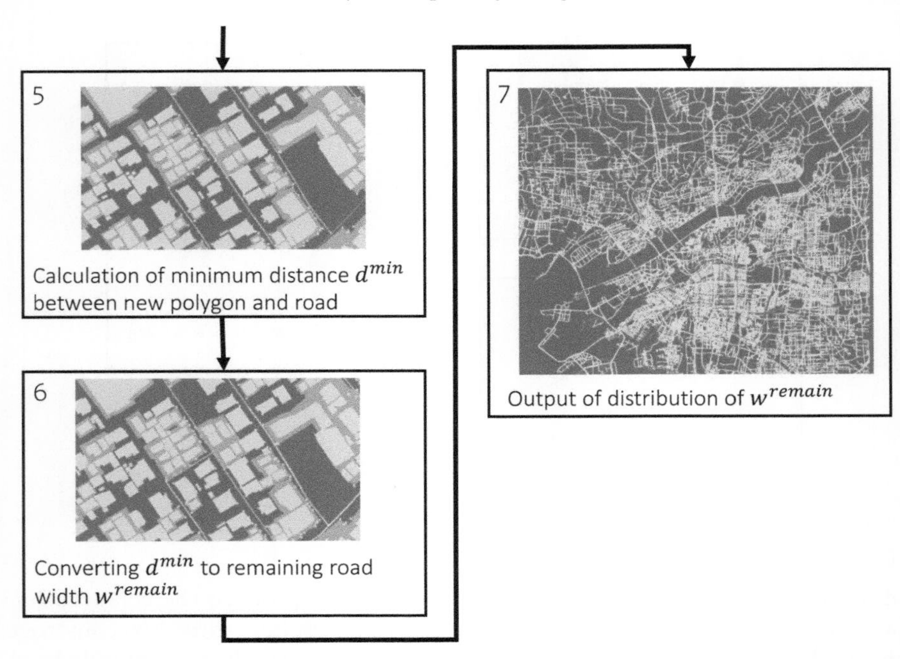

FIGURE 4.8: Procedure of computing road blockage from nearby residential building damage.

for each structure on a road network map; 2) adding information about road nodes, links, and width to each road segment in the map; 3) calculating the average block radius r^b using the maximum drift angle, which is computed by the structural seismic response analysis for each structure; 4) transforming the polygon of the structure to another polygon of the damaged structure using r^b; 5) calculating the minimum distance d^{min} between the polygon of the damaged structure and the road; 6) converting d^{min} to the remaining road width $w^{remaining}$ for each road segment; and 7) updating the road network model using the output of the distribution of d^{min}; $w^{remaining}$ is used to compute the loss of the road capacity, which is given as $1 - w^{remaining}/w$, with w being the original road width. The road capacity is set to 0 if $1 - w^{remaining}/w$ becomes less than 0.5.

4.2.3 Simulation results

FEM enhanced with the HPC capability is used to compute the earthquake waves propagating from a source fault, according to the earthquake scenario of the Cabinet Office, Japan. The distribution of the earthquake wave (which hits the bedrock of the surface ground layer) is computed on the top surface on the model; the top surface is 100×100 km. It should be noted that while the

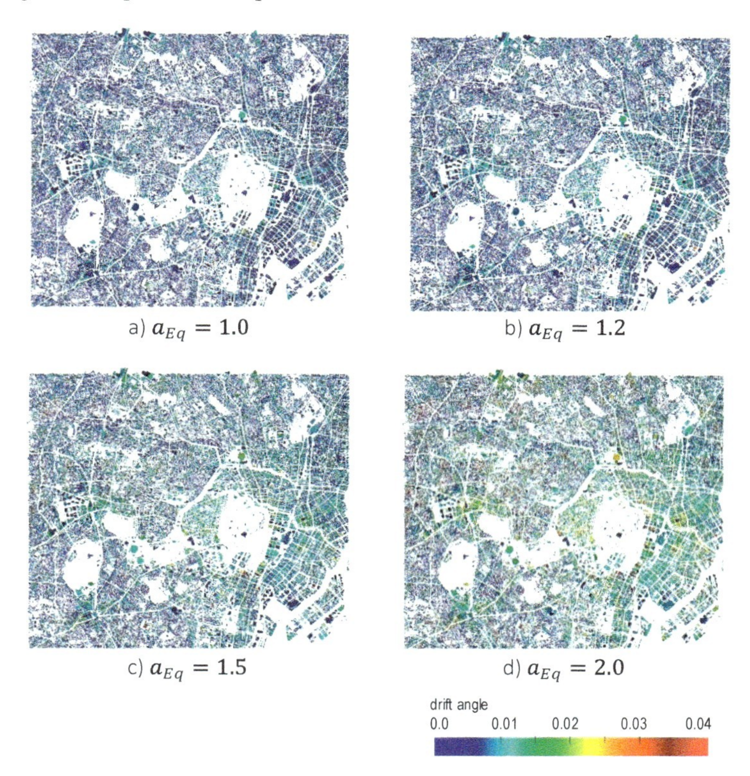

FIGURE 4.9: Distribution of maximum floor-wise drift angle of IES-TME.

element size of the analysis model is of the order of 100 m, we can interpolate the earthquake wave using the interpolate functions of FEM, which are second-order polynomials. Thus, the distribution of the earthquake wave on the top surface is smooth.

The output of the earthquake wave propagation analysis is used as an input of the ground motion amplification analysis. Interpolated earthquake waves are used as the boundary conditions at the bottom of the analysis model of the three-layer ground surfaces. The output of the ground motion amplification analysis is used as the input of the structural seismic response analysis; for each residential building, the ground motion at its site on the surface is used as the input. Soil-structure interaction is not computed[5] in the present IES.

As an example of the simulation results, Fig. 4.9 shows the distribution of the maximum drift angle, denoted by θ^{max}, which is computed for each

[5] It is surely possible to make an analysis model for the structure and soil of each residential building or for a set of few structures and soil. The soil model is extracted from the analysis model of the surface ground in the ground motion amplification analysis.

structure in structural seismic response analysis. To examine the effects of the increase in the ground motion, we change the ground motion, which is computed by the ground motion amplification analysis and is input to the structural seismic response analysis. The change is made by amplifying the time series of ground motion with a constant, denoted by a_{EQ}; the value of a_{EQ} is changed from 0.7 to 2.0. For the case of $a_{EQ} = 1.0$, which is regarded as the reference, the computed value of θ^{max} does not reach 0.01 for a majority of the buildings. This result of the structural seismic response analysis suggests that earthquake damage to structures is not significant for this scenario. As a_{EQ} becomes smaller, θ^{max} becomes smaller, as expected; in the case of $a_{EQ} = 0.7$, θ^{max} does not reach 0.005, which means no damage. As a_{EQ} increases, θ^{max} tends to increase. In the case of $a_{EQ} = 1.5$ there are a few buildings whose θ^{max} exceeds 0.02, which implies severe damage to the buildings. In the case of $a^{EQ} = 2.0$, which means that the input ground motion is doubled, and θ^{max} exceeds 0.02 in various places; there are a few spots in which many structures have $\theta^{max} > 0.02$.

According to the procedures explained in Fig. 4.8, the damages of structures are converted to the damages to the road segment to calculate the loss of road capacity. The distribution of road blockage width in major roads is shown in Fig. 4.10. Here, we consider four cases of changing the input ground

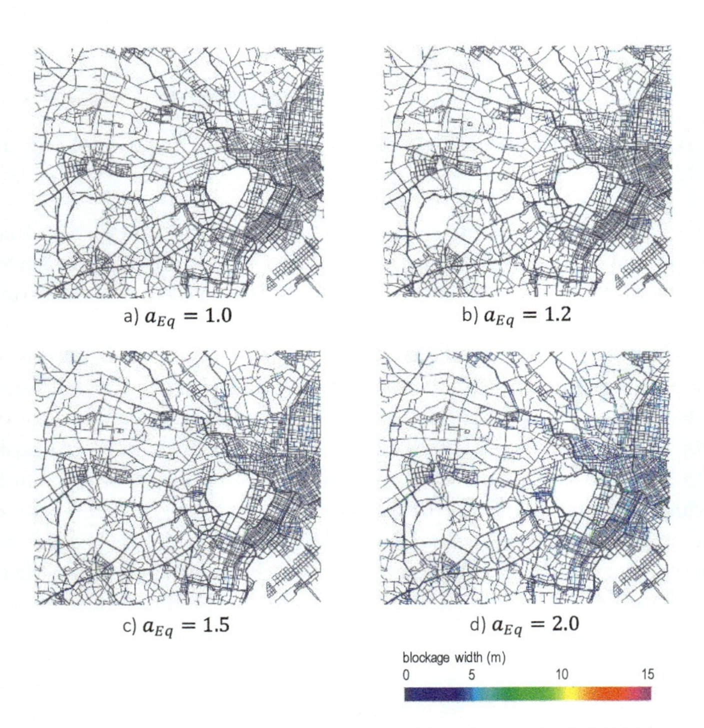

FIGURE 4.10: Distribution of road blockage of IES-TME.

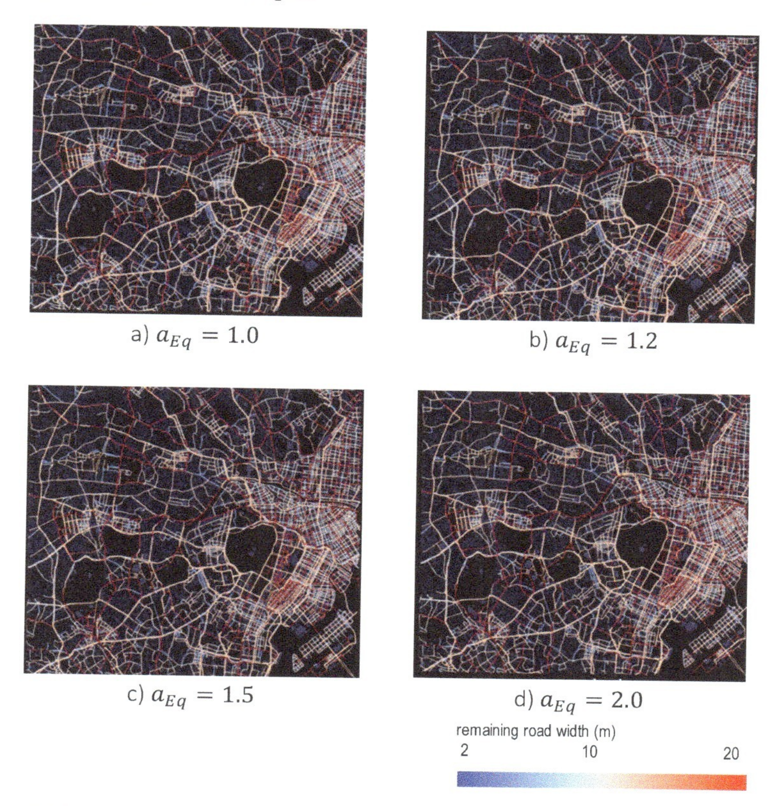

FIGURE 4.11: Distribution of remaining road width of IES-TME.

motion, i.e., $a_{EQ} = 1.0$, 1.2, 1.5 and 2.0. The blockage width is negligible (i.e., less than 1 m) for the case of $a_{EQ} = 1.0$ and 1.2, and there are some roads that have blockage wider than 5 m for the case of $a_{EQ} = 1.5$. Structural damages induce severe road blockage in various places for the case of $a_{EQ} = 2.0$; almost all road segments in the east part have road blockage wider than 10 m. The blockage of a road segment is determined by the relative location of the nearby structures as well as their damage degree, which is influenced by ground motions at their site. It is necessary to carry out integrated simulation such as IES for an accurate estimate of road blockage. In Fig. 4.11, the distribution of $w^{remaining}$ is plotted for the four cases of a_{EQ}. The east part of Tokyo appears to be seriously damaged, as shown in Fig. 4.10, which shows the road blockage width[6] for the four cases of a_{EQ}. However, if the remaining road width, $w^{remaining}$, is plotted instead of the road blockage width, the

[6]The road blockage width is used as an index of the effects of the structural damage on the road network, and the remaining road width is used to determine the road capacity in the traffic simulation.

damages to the road network are not very clear[7] in Fig. 4.11 for all the four cases of a_{EQ}.

Using the updated models of the damaged road network, we carry out the traffic flow simulation. In general, greater uncertainties are involved in the traffic flow simulation for the initial conditions of the traffic state, and it is difficult to make an accurate estimate of detailed traffic flow after an earthquake. For better preparedness, it is important to find the points at which traffic congestions would be induced. Considering many earthquake scenarios and finding points of traffic congestion is a way to account for the uncertainties in finding such points in the road network.

As an example of the numerical results of the traffic flow simulation, Fig. 4.12 shows that the distribution of the relative changes in the traffic density and the traffic congestions, compared with the ordinary traffic state. Two cases of earthquake damage on the road network are used; Case 2 uses a more seriously damaged road network model than Case 1; while damages are concentrated in the east and central part of the road network in Case 1, larger parts of the road network from the east to the west have damaged road segments in Case 2. As is seen, little difference can be observed for the traffic density shown in Fig. 4.12a). In terms of the traffic congestion, we can observe a slightly larger reduction in the traffic flow in Fig. 4.12b). However, compared with the road network damage, the traffic congestion induced by the damage appears to be less severe.

It is certainly true that the traffic density and congestions would be different if other traffic demands are input into the traffic flow simulation. In general, the effects of the road network damage on the traffic flow become larger as the traffic demand increases. It is thus necessary to consider various initial conditions as well as various traffic demands for traffic flow simulation. There are same points in the road network at which larger traffic congestion happens, as shown in Fig 4.12b). As mentioned, higher priority should be put on these points, considering a retrofitting plan of road structures, even though considerable uncertainties are inevitably included in the current traffic simulation.

As explained in Chapter 1, the estimate of earthquake hazard and disaster ought to be made using an un-biased method, which refers to the correct estimate of the mean; according to the law of large numbers, the sum of all un-biased estimates will be accurate even though each estimate is not accurate. It is difficult to make the Earth science simulation (the earthquake wave propagation analysis) un-biased because the correctness of the simulated earthquake waves depends on the quality of the analysis model of the crust, even if numerous earthquake scenarios are used in the simulation. However, as for earthquake engineering simulation (the structural seismic response

[7] It could be regarded as misleading due to the visualization. Visualizing the analysis results in high resolution is intuitively understandable but the same results can yield different understandings, as shown in Figs. 4.10 and 4.11.

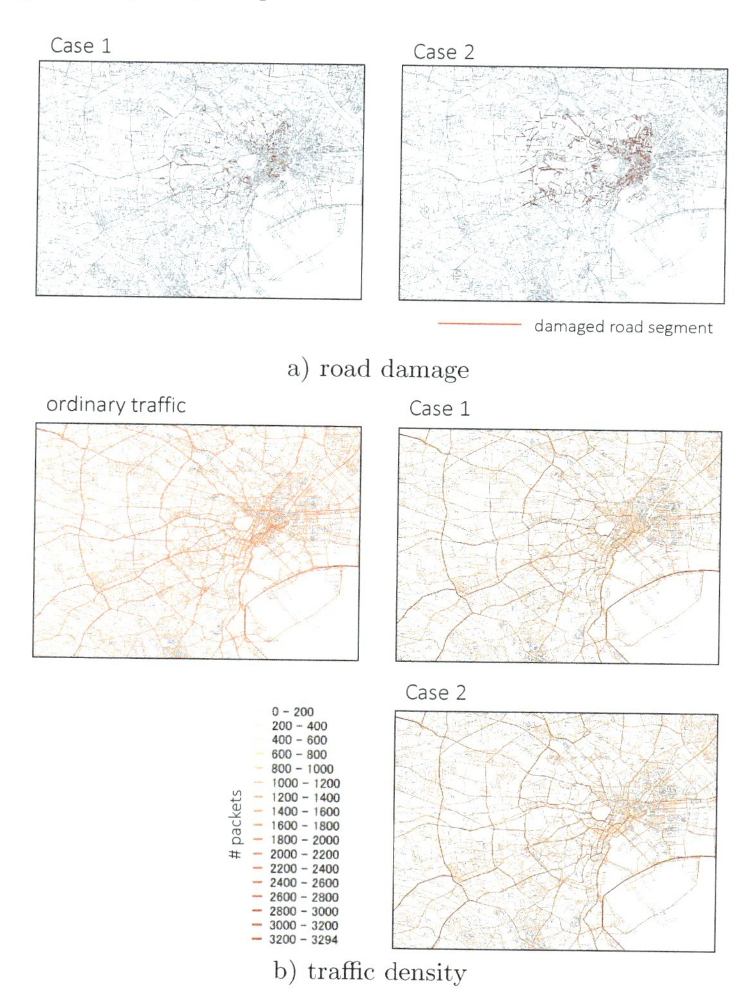

a) road damage

b) traffic density

analysis), we can consider that the simulation is un-biased if the construction of an analysis models for each structure is made in an un-biased manner. While it is not possible to examine whether the construction of the analysis models are un-biased, we can expect that the method of constructing analysis models from the data about the configuration and structural property does not include any process which induces a bias of large degree in the constructed analysis model. Therefore, while the structural damage estimate made for each residential building in IES-TME is not always correct, the sum of the structural damage estimates taken in a certain area becomes more accurate.

It is not straightforward to apply the law of large numbers to the traffic flow simulation conducted in IES-TME. Thus, we cannot say that even if the

c) traffic congestion

FIGURE 4.12: Traffic conditions of IES-TME.

estimate of local traffic congestion is not always correct, the sum of such local estimate will be accurate. This is because traffic flow simulation analyzes the movement of all the vehicles on the road network and local congestions influence the traffic flow in other parts. The interaction between the neighboring structures is usually ignored in earthquake engineering; hence, seismic structural analysis can be separately made for each structure. Separating the road network model is not accepted in traffic flow simulation because the results are not the same if the road network models are separated. We need to study the reliability of the traffic flow simulation to assure that the estimates of traffic flow in a certain part of the road network are accurate even though the local estimate may not be always correct.

4.3 Nankai Trough Earthquake

In this section, we simulate the Nankai Trough Earthquake, which is considered to hit densely populated and highly industrialized regions of Japan. Like the Tokyo Metropolis Earthquake, several estimates have been made using empirical methods in this large-scale earthquake. We have arranged the first three subsections in the same manner as the previous two examples. We have

TABLE 4.2: Problem setting of IES-NTE.

Earth Science	Earthquake scenario	3 scenarios given by Cabinet Office Prediction
Earthquake engineering	Surface ground	Osaka area: synthesized ground motion other area: ground motion at bedrock predicted by Cabinet Office
	Structures	1,266,706 multi-degree-of-freedom models
Social science	Traffic flow	495,595 links
		558,572 packets of 15 vehicles
	Traffic demand	Use of data of 2017 Kumamoto Earthquake
	Economic activity	Osaka, Kansai and whole Japan, 4,320 enterprises of 36 industry sectors, household to foreign industry

added a subsection for mass evacuation simulation in this section, which will be a key element to minimize tsunami disaster induced by the Nankai Trough Earthquake.

4.3.1 Problem setting

The Nankai Trough Earthquake is an inter-plate earthquake, and it influences a larger region than the Tokyo Metropolis Earthquake, in which an intra-plate earthquake is used as the earthquake scenario in the preceding subsection. We carry out IES for the Nankai Trough Earthquake (IES-NTE) considering the same problem setting of IES-TME but increasing the scale of the problems; for instance, the distribution of ground motion is computed for the Kansai Region, which is a more than 10 times larger area than the Tokyo Metropolis. This problem setting is summarized in Table 4.2; in addition to the problem scale, IES-NTE has the following two new features: 1) the number of earthquake scenarios is increased to three; and 2) the traffic demand simulation is added in the traffic simulation.

We consider three earthquake scenarios studied by the Cabinet Office, Japan; see Cabinet Office, Japan (2021b). These scenarios are called basic, east side, and land side, hereinafter; the basic scenario[8] is regarded as an earthquake scenario that is used to consider retrofitting plans or other disaster mitigation actions; the east side scenario is an earthquake scenario in which the east side of the fault moves on the plate boundary; and the land side scenario is the worst earthquake scenario as its epicenter is close to the Japanese Islands and generates strongest ground motions in the Kansai Region. Using supercomputers, we cannot conduct Earth science simulation of the earthquake wave propagation at high temporal and spatial resolutions so that the results of this simulation for earthquake hazard are used as the input of earthquake engineering simulation for the estimation of earthquake disaster. FEM enhanced with HPC capability and an analysis model of the crust

[8]It is taken for granted that the probability of the basic scenario is larger than that of the other two scenarios.

structure of the region that is influenced by Nankai Trough Earthquake are available. However, the simulation cannot[9] be performed due to the limitation of the current computing power.

As for ground motion amplification, it is possible to analyze Osaka City using FEM enhanced with HPC capability. An analysis model of the same scale and fidelity as the one used in IES-TME must be constructed; the ground motion amplification processes can be computed for other areas using a suitable analysis model. However, such an analysis model is not available. As an alternative simulation of FEM analysis, we use a program for analyzing the ground motion amplification processes in a stratified underground structure for each site of a structure; see Iiyama *et al.* (2019). While the topographical effects[10] of three-dimensional underground structures are ignored in the alternative simulation, it is readily applied to any site of the structure because it is easy to construct a stratified underground model. As for the structural seismic response analysis, residential buildings and road bridges are computed. Like IES-TME, the residential buildings include high rises in the business districts and shopping and entertainment complexes, in addition to residential houses with one or a few stories. A simple model for structural seismic response analysis is made for road bridges, the damages of which can shut down the traffic flow nearby.

The traffic demand and flow simulations are carried out in IES-TNE. As mentioned in the preceding subsection, the estimate of traffic demand after an earthquake is more important than considered; hence, the traffic demand simulation is added in IES-NTE. Post-disaster traffic demand changes from the ordinary one, depending on the degree of earthquake disaster. Hence, it is important to analyze this demand considering earthquake disaster. The traffic flow simulation is conducted for the entire Kansai Region; while the area is wider than Central Metropolis Region, the road network is of the same scale as the one used in IES-TME due to the smaller density of the road network.

4.3.2 Constructed analysis models

As mentioned, FEM enhanced with HPC capability is not applied to analyze the earthquake wave propagation processes in IES-NTE, even though an analysis model of the crust structure is readily constructed for the Kansai Region. Indeed, an analysis model of the crust structure for the east part of Japan is constructed, and the analysis model is used for the quasi-static analysis

---

[9]If the analysis model is decomposed into a few small analysis models, the current supercomputer can execute FEM with HPC capability in each small analysis model. This simulation must consider the coupling among these small analysis models by suitably setting the boundary conditions on the interfaces of the neighboring analysis models.

[10]While the topographical effects are not considered, the stratified underground model of the surface ground layer can be used to include the effects of soil-structure interaction in structural seismic response analysis.

TABLE 4.3: Model parameters of surface ground of IES-NTE.

	Thickness [m]	V_s [m/s]	V_P [m/s]	Density [kN/m^3]	Reference strain [μ]	Damping factor [%]
Surface	30	AVS30	Computed from V_S	20	10	3.0
Bedrock	50	700	2100	20.5	10	0.5

of the crust deformation, which does not require high spatial resolution that is needed for the dynamic analysis of the earthquake wave propagation processes. Data regarding the configuration and material property of the crust structures, which are based on seismic monitoring on land and the ocean bottom, are used to construct the analysis model of the 2011 Great East Japan Earthquake. An analysis model of the Kansai Region will be constructed using similar data.

In the Kansai Region of the present example, there exist 1,266,706 residential buildings. An analysis model of stratified surface ground layers is constructed for the ground at each residential building site, and a linear multi-degree-of-freedom model is constructed for each residential building. These models are used for the ground amplification analysis and structural seismic response analysis, respectively. The stratified underground model consists of two layers, a bedrock of 50 m thickness and $V_s = 700$ m/s and a surface ground layer of 30 m thickness. The shear wave velocity of the surface layer is determined using the data[11] provided by the National Research Institute for Earth Science and Disaster Resilience (NIED), Japan. The stratified underground model employs a nonlinear RO model for soil. Table 4.3 summarizes the model parameters of the surface ground layers. The multi-degree-of-freedom models are used by combining the 3D map and land registration, which is available in Osaka City; see the automated model construction explained in the preceding section. The residential buildings are classified as wooden, steel, reinforced concrete, or steel-reinforced concrete building, and the model parameters of the multi-degree-of-freedom model are determined accordingly.

Automated model construction is applied for the road bridges, which are located in the Kansai Region in this example. Data about road bridges are scattered, and very little information is available about the structural properties; it is not even possible to identify a road segment in which a bridge is constructed. As shown in Fig. 4.13, we consider the procedures of automatically constructing the bridge models, which is a simplification of the automated model construction of the road bridges explained in the preceding section.

For traffic demand simulation, we need to analyze the earthquake damages to various structures besides the residential buildings and road bridges. In the present example, physics-based simulation is not applicable to these structures. For infrastructures such as lifelines, we use a conventional method

[11]The data used are AVS30 (average shear-wave velocity of the ground in the upper 30 m depth.

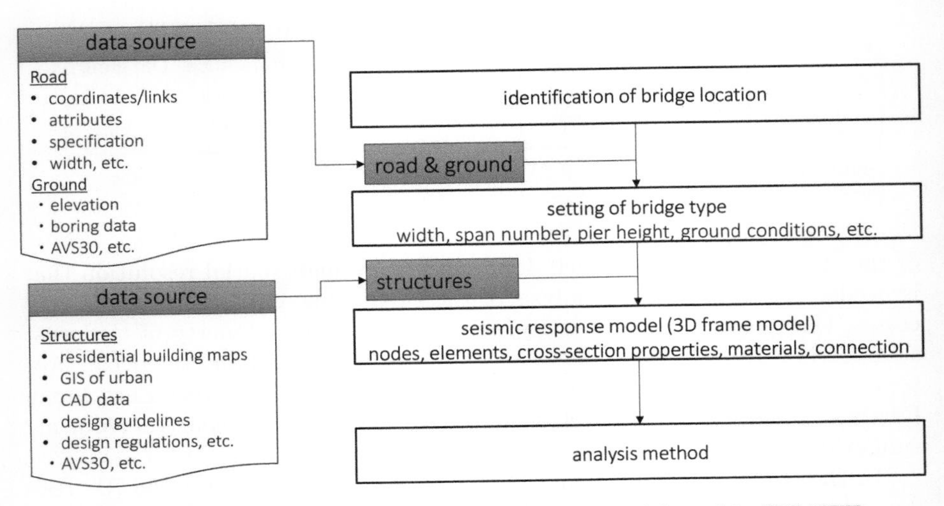

FIGURE 4.13: Procedure of constructing bridge model used in IES-NTE.

of using fragility curves; the ground motion simulated at the site of a structure is used instead of an attenuation curve. The structures analyzed using the fragility curve are buried pipelines of sewerage, water supply, electricity, and city gas as well as port facilities. In addition to the infrastructural damages, traffic demand simulation uses the following four input data about the evacuees: 1) area-wise population given by the Japanese Government; 2) the locations of evacuation shelters provided by Osaka City; 3) information about retailers given by the 2014 commercial statistics of Ministry of Economy, Trade and Industry, Japan; and 4) the locations of public baths[12] given by Osaka Public Bath Unition. Traffic demand is generated for each resident in a damaged area who needs to act by moving to a suitable shelter, receiving emergency foods and supplies, and going to a nearby public bath. Data about the resident behaviors in the 2016 Kumamoto Earthquake are used to determine the probability of these actions for each evacuee.

As for the traffic flow simulation, we analyze the entire Kansai Region which is larger than the Tokyo Region; this region covers the Osaka, Kyoto, Hyogo, Nara, Wakayama, and Okayama prefectures. For the road network of this region, we construct a road network model. The node and link numbers of this model are 183,160 and 434,678, respectively; as mentioned, this model is of the same scale as the model used in Central Metropolis Region of IES-TME. Traffic demand data are given as the traffic volume for each traffic which is specified by the pair of origin and destination. The time frame for the traffic flow simulation is in the morning from 4:00 a.m. to noon.

[12]The use of public baths is very important in Japan during emergency situations.

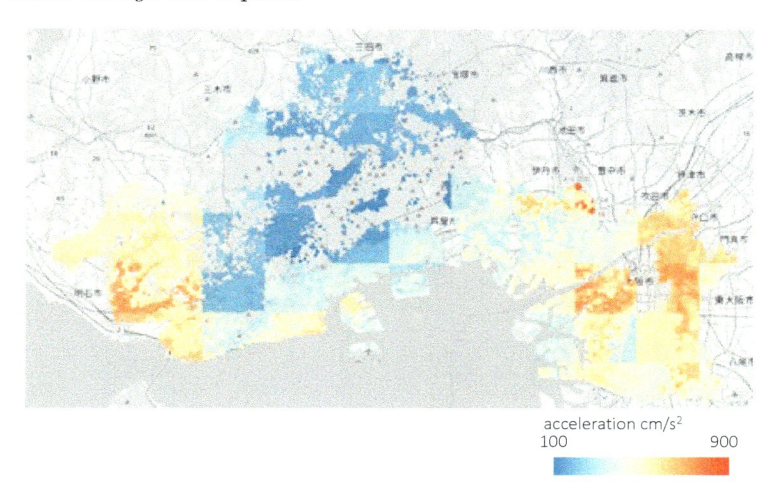

FIGURE 4.14: Distribution of maximum ground acceleration of IES-NTE.

TABLE 4.4: Average and maximum ground motion computed of IES-NTE.

Scenarios		G^{ave} (cm/s^2)	G^{max} (cm/s^2)
Basic		348.9	4502.2
East side		229.0	3222.3
Land side	x 1.0	174.7	2688.9
	x 1.5	325.2	4702.3
	V_s 50% increase	344.0	4680.5

4.3.3 Simulation results

In Fig. 4.14, the distribution of maximum ground motion for the basic scenario is plotted. As mentioned, in this example, Earth science simulation is not carried out, and earthquake waves given by the Cabinet Office, Japan are used as the input of the ground motion amplification processes. Each tile in this figure shares the same seismic wave and the difference in the underground structure causes a difference in the ground motion[13] at the surface. Table 4.4 summarizes the average and maximum ground motion computed in the present example. As for the land-side scenario, the following two cases are added: 1) the earthquake wave is magnified by 1.5; and 2) V_s of the ground surface

[13]Since the coincidence of the natural frequency of the surface ground layer and the structure leads to larger seismic response, a simple analysis model of the surface ground layers is useful in the seismic response analysis, provided that the natural frequency is accurately computed using the simple model.

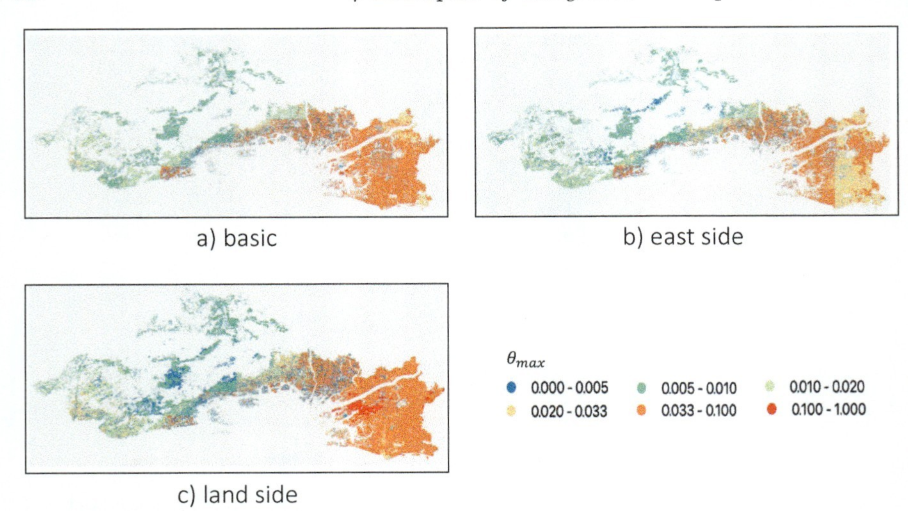

FIGURE 4.15: Distribution of maximum drift angle of IES-NTE.

is increased 50%. Table 4.4 summarizes the average and maximum ground motion[14] computed for these earthquake scenarios.

In Fig. 4.15, the distribution of the maximum drift angle, θ_{max}, is plotted for the three earthquake scenarios. According to the basic scenario, Osaka City in the east part is severely damaged as the maximum drift angle[15] exceeds 0.03. The east and land side scenarios yield similarly severe damages to Osaka City, although θ_{max} becomes less than 0.02 in the east south areas according to the east side scenario. The west part is less severely damaged in all the three scenarios. This is mainly due to the difference in the input ground motion at the bedrock; see Fig. 4.14. The ground motion is less strong in the west part, and this difference yields a high contrast of the structural damages in the east and west parts. There are some places where θ_{max} exceeds 0.02 in the west side according to the three scenarios. As mentioned, these places should have a higher priority in a mitigation plan.

Like IES-TME, the structural damages are converted to the reduction of the road capacity to construct an analysis model of the damaged road network. An example of the conversion that takes advantage of the numerical results of high resolution is shown in Fig. 4.16. A structure-wise damage is used to compute the road blockage, which is converted to the reduction of the road width in a road segment. The reduced road width reduces the road capacity

[14]It should be pointed out that values of the maximum ground motion shown in Table 4.4 are not realistic. Careful examinations are needed to find out the causes of such large values for the maximum acceleration.

[15]In the present structural seismic response analysis of the multi-degree-of-freedom models, the structure is severely damaged or partially collapsed if the maximum drift angle exceeds 0.03.

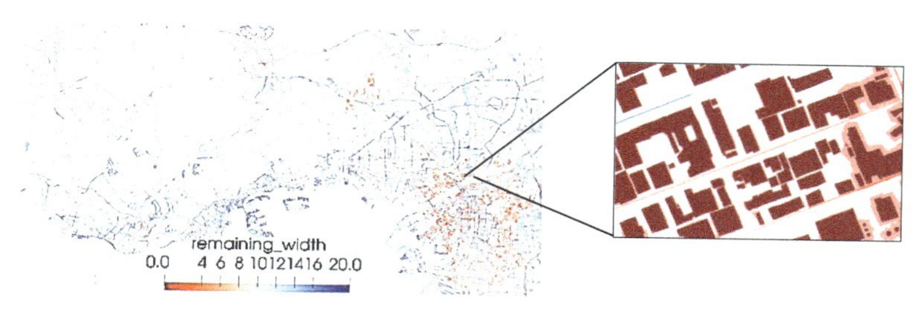

FIGURE 4.16: Distribution of remaining road width of IES-NTE with close-up view of road network.

of a road segment, and the road network model, which is updated using the reduced road capacity, is used in the traffic flow simulation.

The traffic demand and flow simulations are carried out in Osaka City, using the damaged road network model which is constructed on the basis of the physics-based simulation and the fragility curves of the infrastructures. In Fig. 4.17, the traffic demand, which is measured in terms of the number of travelers, is presented; a) is the distribution of the ordinary traffic demand, and b), c), and d) are the distribution according to the basic, east side, and land side scenarios, respectively. In general, traffic demands are reduced based on the degree of the earthquake damages. In the north part of Osaka City, which is severely damaged, the traffic demand is significantly reduced. In the west part, which is less severely damaged, the reduction in the traffic demand is significant as well; residents in this area commute to Osaka City, and this traffic demand of commuting is reduced.

Inputting the traffic demand explained in the above to the damaged road network model, we carry out the traffic flow simulation. As an illustrative example, the distribution of traffic congestion is presented in Fig. 4.18; similar to Fig. 4.17, a) is the ordinary traffic congestion, and b), c), and d) are the traffic congestions according to the basic, east side, and land side scenarios, respectively. While IES-TME does not produce significant traffic congestion, IES-NTE yields a heavy traffic congestion in Osaka City. According to the basic scenario, the congestion is spread through the road network from Osaka City to the west side of the Kansai Region. The east and land side scenarios yield the same distribution of the traffic congestions. It should be noted that according to the land side scenario, the degree of traffic congestion in the Kansai Region is the greatest, and traffic congestions take place even in the north side of Osaka City.

An advantage of IES is that the processes of earthquake hazard, disaster, and response can be observed in a backward manner. In view of Fig. 4.18, we notice that a large-scale traffic congestion occurs in Osaka City and its surrounding cities. Since the traffic demand is not large in these area, as shown in Fig. 4.17, the traffic congestion is created by physical damages. Indeed, as shown in Fig. 4.15, these areas suffer from severe damages. To reduce the

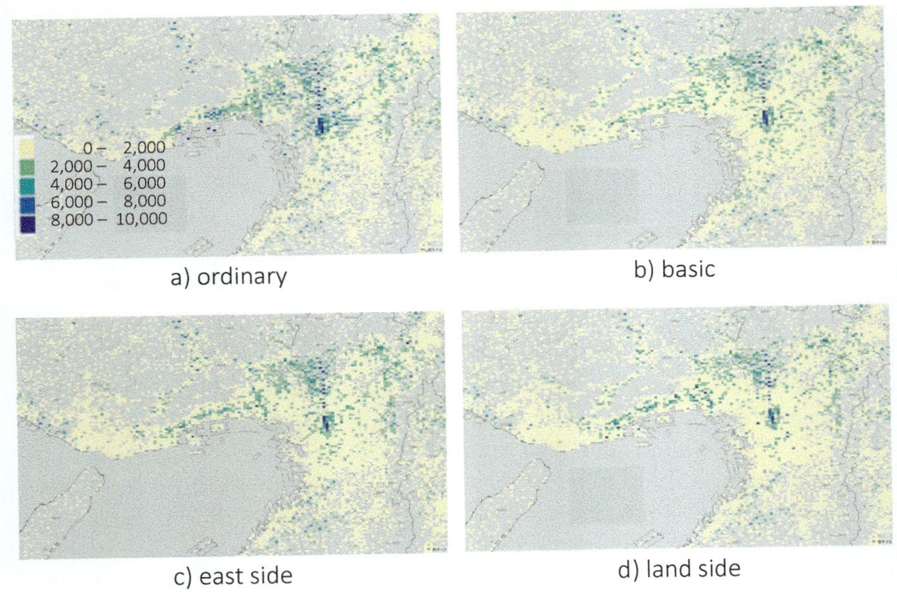

FIGURE 4.17: Distribution of traffic demand of IES-NTE.

traffic congestion, we can examine the required degree of retrofitting of the structures so that the road network damages are reduced. In the conventional forward manner, it is not easy to examine the requirement of the retrofitting structures which result in the traffic congestion. This backward visualization is an alternative since it first identifies[16] the location of heavy traffic congestion and then the areas that cause the traffic congestion. It should be noted that a severely damaged area does not always cause traffic congestions.

While IES-TME uses a massive solid element model for the surface ground layers, the present IES-NTE uses site-wise stratified underground model. The solid element model has an advantage of computing the topographical effects accurately, but it is not certain that the results of this model area are not biased. The simple stratified underground model is not biased as the simple multi-degree-of-freedom model for a residential building as these models are constructed from the actual data of the ground and the structures. Thus, according to the law of large numbers, we can expect that the stratified underground models are not always correct, but the sum of the estimates made using these models will be accurate. If the quality and quantity of digital data about the underground structure are limited, it is acceptable to use site-wise stratified underground models, even though these models cannot compute the topographical effects like a solid element analysis model of high fidelity.

[16]The conventional forward manner first identifies areas where ground motions are greater or areas where structures are severely damaged.

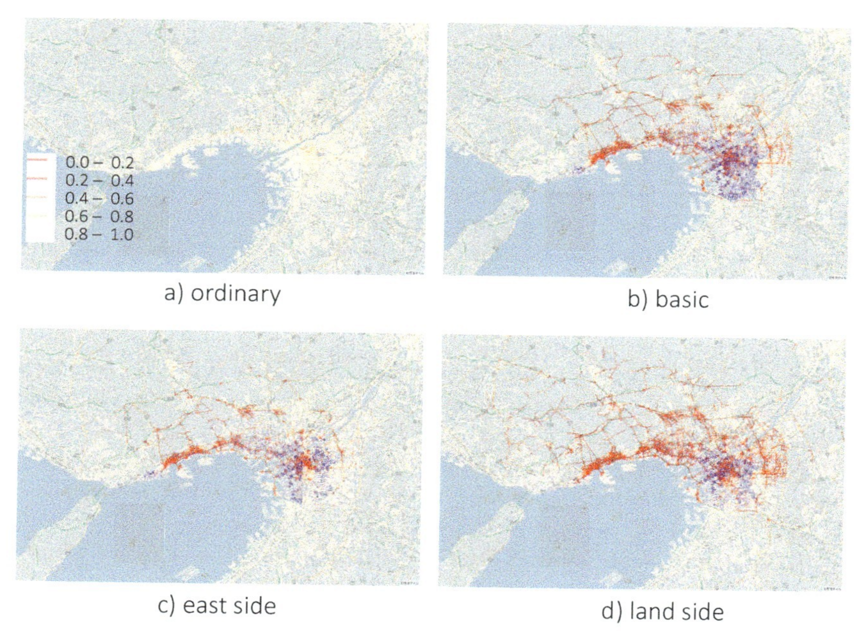

FIGURE 4.18: Distribution of traffic congestion of IES-NTE.

Unlike IES-TME, the occurrence of serious traffic congestions is observed in IES-NTE. The coupling of the traffic demand simulation and traffic flow simulation contributes to it, while severe earthquake disaster is computed in the earthquake engineering simulation. The reliability of traffic simulation is not examined as discussed in the example of IES-TME; the law of large numbers is applicable to the traffic demand simulation if it does not consider interaction among the evacuees. We need to study the validation[17] of the analysis models used in IES, examining that the results obtained by using the analysis model are not biased.

4.3.4 Mass evacuation simulation

In this subsection, simulations of some tsunami-triggered mass evacuation scenarios are presented as demonstrative applications of the developed agent-based evacuation simulator. As discussed in Section 2.3, most simulators model mass evacuation as agent flows on 1D networks and realistic evacuee flow rates are attained by setting the agents' speeds according to the average density in each link and a set of suitable fundamental diagrams. On the contrary, as explained in Section 2.3.4, the parameters of the developed autonomous agents are tuned such that the agents' individual actions reproduce characteristics

[17]The validation that examines the bias (i.e., the mean is exact) of the numerical results is practical for social simulation.

such as fundamental diagrams as an emergent behavior (Leonel *et al.* 2013 and Leonel *et al.* 2014). This ability of the agents combined with other constituent functions of the agents such as g^{see}, $g^{identify\_env}$, and $g^{is\_path\_blocked}$, and the detailed model of the environment make it possible to simulate congestions on debris-scattered roads, interactions of pedestrians and cars at non-signalized junctions, etc. Simple agent-based models are not applicable in such scenarios, for which prior real-world observations are unavailable and which requires the best use of traversable space, complicated agents' interactions, etc. Some of the presented hypothetical scenarios involve conditions such as debris scattered over open spaces and wide roads, inundation, and non-signalized junctions to demonstrate the advantages of the developed agent-based model in simulating demanding scenarios. The presented scenarios can be categorized into two main groups, namely, involving only the pedestrian mode and the car-pedestrian mixed mode.

All the demonstrative simulations presented in this subsection use hypothetical problem settings; hence, the simulation results do not reflect the real problems involved in the evacuation of the simulated area.

This section presents several pedestrian mode scenarios, some of which include the effects of fallen debris, inundation, and visibility limitations. The last scenario, which involves nighttime evacuation during a festival event, highlights the need of a detailed model of the environment and sophisticated agents.

Problem settings

We used a 8.5 km×5.4 km region of Kochi city located on Shikoku island, Japan. According to the population of the area, 57,000 agents are randomly scattered close to the buildings. The agents are categorized into two groups, as shown in Table 4.5. The regions above 30 m elevation, shown in light gray, are considered as the only available evacuation areas. This is a highly exaggerated hypothetical scenario compared to the maximum tsunami heights expected in Kochi city area.

Earthquake-induced damages are estimated using physics-based seismic response analysis simulators, as explained in Section 2.2 using the strong ground motion of the 1995 Kobe earthquake. Buildings are deemed to be damaged if

TABLE 4.5: Properties of the two age groups of agents.

Age	Below 50 years	Above 50 years
Percentage	55	45
Speed (m/s)	1.43 ± 0.11	1.39 ± 0.19 4
Distance of eye sight (m)	50	50
Pre-evacuation time (s)	1000 ± 600	1000 ± 600

the inter-story drift angle is larger than 0.005 (Theodore *et al.* 1986), which is a simple criterion often used in earthquake engineering. Footprints of damaged buildings are increased by 40% of their height (Xue *et al.* 2012) to model the scattering of debris. The arrival time of the tsunami is set to 15 minutes and the grid is updated every 3 minutes, according to the inundation data provided by Takashi *et al.* (2014), to mimic tsunami inundation.

Monte-Carlo simulations

Monte-Carlo simulations are conducted in order to increase the reliability of the simulation results by taking the effects of the involved uncertainties into account. The only random variables considered in the presented simulations are the distribution of evacuees and their speeds. To decide a sufficient number of simulations required for each Monte-Carlo simulation, 1000 simulations with random initial distributions and speeds are conducted, and the convergence of standard deviation of the total number of agents evacuated within 40 minutes is analyzed as a global measure, while the number of agents evacuated at every 30 seconds is analyzed as a local measure. The scenario considered is evacuation to high grounds during an ordinary day (i.e., without any earthquake or tsunami inundation). As seen in Fig. 4.19a, the standard deviation of the total number evacuated has a negligibly low variation, indicating global convergence. In addition, Fig. 4.19b shows that the mean and standard deviations of 400 simulation samples is nearly identical to those of 1000 samples, indicating the convergence of this local measure. Since both of the above considered global and local measures converge, 400 simulation samples are considered to be sufficient for the Monte-Carlo simulations presented in this section. Further, Fig. 4.19c indicates that the standard deviation of the cumulative number evacuated is negligibly small. Therefore, only the mean of the number evacuated is shown in the presented results. 400 is a reasonable sample size, considering the facts that we have only the location and speed as random variables, the number of agents is large, and the standard deviation of speed is small. However, real applications may need much larger number of samples depending on the nature of the random variables involved.

Day time evacuation

We consider four daytime evacuation scenarios: ordinary conditions (i.e., without any damage to the environment or inundation), with earthquake disaster, with tsunami inundation, and with both the earthquake and tsunami. In all these scenarios, only 57,000 resident agents are considered and their initial positions are randomly set within 20 m proximity of the buildings.

Figure 4.20 shows the time histories of the cumulative number of agents evacuated for each of the four scenarios. As is seen, both earthquake damage and tsunami inundation reduce the number of evacuees almost by the same amount. However, while the effects of earthquake damage start to appear at early stages, the effect of tsunami inundation appears after 20 minutes since

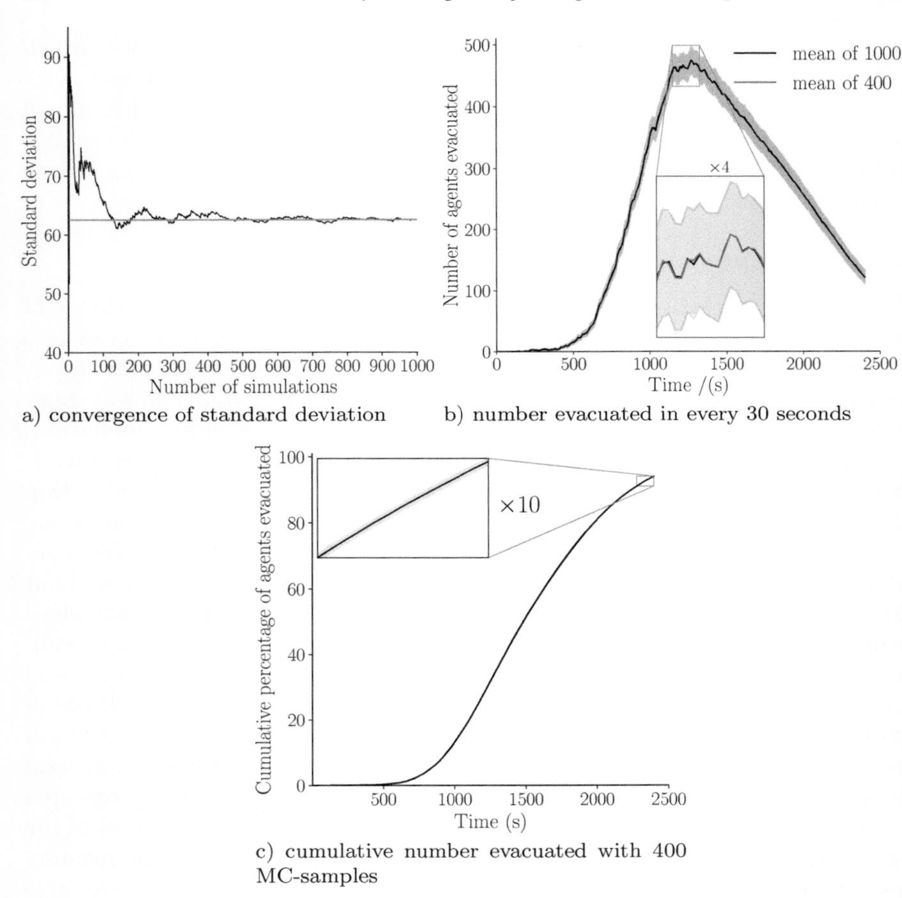

a) convergence of standard deviation

b) number evacuated in every 30 seconds

c) cumulative number evacuated with 400 MC-samples

FIGURE 4.19: Convergence of the standard deviation and the cumulative number of agents evacuated with the number of samples in Monte-Carlo (MC) simulations. Shown in light gray is one standard deviation.

the arrival time of the tsunami is set to 15 minutes. At the end of the 40 minutes period, we observe a 13% reduction in the total number evacuated due to inundation, 15% reduction due to earthquake, and 26% due to both the earthquake and inundation.

The comparison of road usages between the scenario with earthquake damage and the scenario with tsunami inundation indicated that tsunami inundation significantly reduced the number of agents crossing the three bridges, highlighted with black ellipses in Fig. 4.21a. Closer investigations showed that the tsunami made these three bridges inaccessible due to the inundation of their access roads. In order to further investigate whether this cut-off of the bridges plays a central role in reducing the number of evacuees, another scenario is considered with these bridges completely blocked from the beginning.

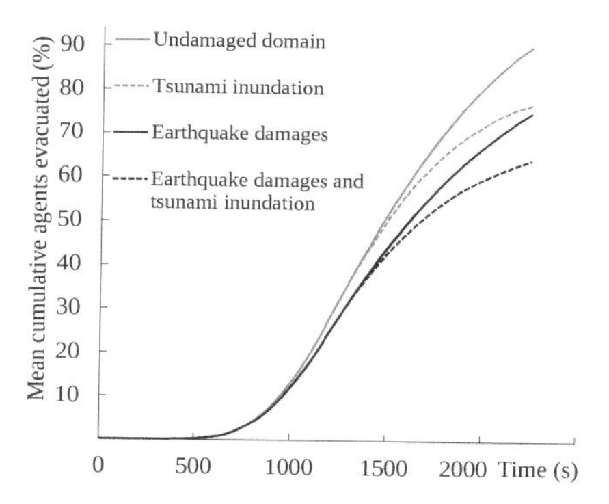

FIGURE 4.20: Evacuation time histories under different conditions.

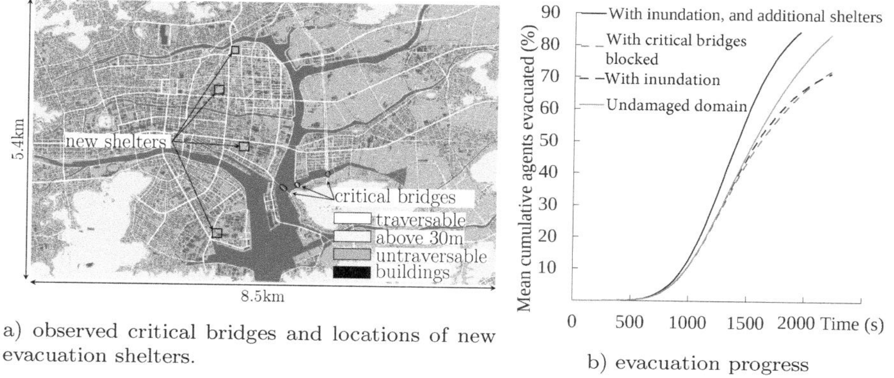

a) observed critical bridges and locations of new evacuation shelters.

b) evacuation progress

FIGURE 4.21: Evaluation of the importance of three critical bridges and advantages of new evacuation shelters.

As seen in Fig. 4.21b, the cut-off of these bridges indeed significantly reduces the number of evacuated agents, indicating that securing access to these bridges can significantly improve the progress of evacuation. As a strategy to address the cut-off of the three bridges, another scenario is simulated by introducing large evacuation shelters closer to the three bridges and the other two main bridges; see Fig. 4.21b. As seen in Fig. 4.21b, this strategy not only improves the evacuation progress by 19% by the end of 33 minutes but also leads to about 12% improvement compared to the scenario without any earthquake or tsunami.

Nighttime evacuation under low-lighting conditions during festival

In order to highlight the need of a high-resolution model of the environment and autonomous agents, nighttime evacuation during a festival occasion is presented. This scenario considers a full moon night, in which an earthquake

TABLE 4.6: Speed of pedestrians under different lighting conditions.

		Age	
Light conditions	Visibility	Below 50 years	Above 50 years
0.2 lux	15m	70%	50%
15 lux	30m	90.6%	83%

has damaged the environment, causing a power failure, and a tsunami is expected to arrive in 15 minutes after the earthquake. 18,000 visitors and 18,000 residents are assumed to be participating in the festival, which takes place in a 14 km^2 rectangular area shown in Fig. 4.22a. The agents participating in the festival are randomly distributed across the streets and open spaces, while another 39,000 residents are distributed over the whole domain. At full moon, 0.2 lux of lighting and 15 m sight distance are assumed (Table 4.6). The visitor agents are considered to be unfamiliar of the surroundings and their only means of evacuation is set to follow the residents in their vicinity. The low-lighting condition makes it difficult for them to find and follow the resident agents; sharp turns and scattered debris can make the visitors lose sight of their followee. A second scenario with emergency street lighting of 15 lux at 30 m spacing, which is equal to common street lighting (Ouellette *et al.* 1989), is considered as a mitigation measure. This emergency lighting enables 30 m visibility. The maximum speeds of the agents under these lighting conditions are set according to Table 4.7. Further, it is assumed that evacuees prefer to take safer paths, which were estimated using Algorithm 3.

According to the cumulative number of the evacuated shown in Fig. 4.22b, the low-lighting condition significantly reduces the number of agents evacuated, compared to that on an ordinary day. Further, it is observed that providing lighting of 15 lux significantly enhances the ability of the visitor agents

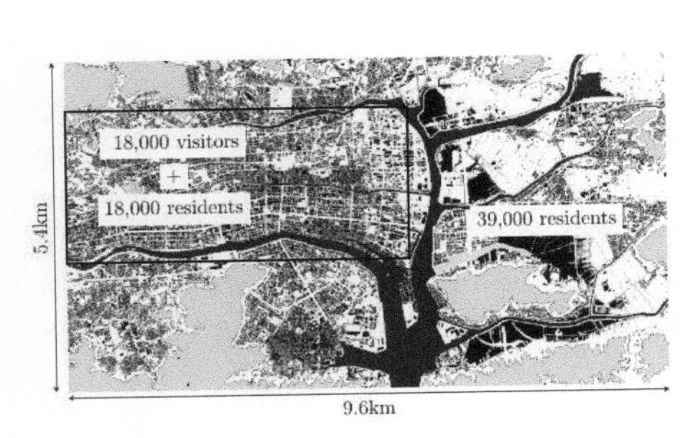

a) distribution of agents during the festival event

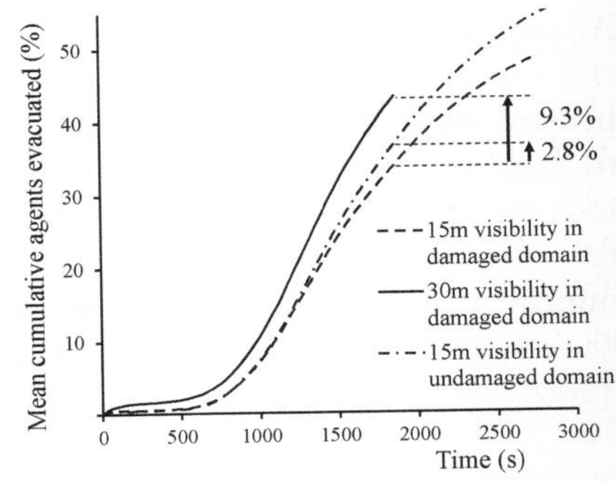

b) cumulative number of agents evacuated

FIGURE 4.22: Effects of evacuation under different lighting conditions during a festival occasion.

TABLE 4.7: Speed and evacuation start time for the car+pedestrian mode scenario. The start time of the cars are capped to exclude values below 200 s and above 1200 s. The abbreviation S.D. stands for standard deviation.

Type	Pedestrian		Car	
	Mean	S.D.	Mean	S.D.
Speed (m/s)	1.28	0.14	11	1
Start time (s)	1500	240	710	220

to locate and follow others. An additional scenario with moonlight (i.e., 15 m visibility) without the earthquake-induced damage is simulated as a comparative estimation of the effects of the lighting conditions. As shown in Fig. 4.22b, the influence of the visibility conditions is significantly high, at least for the particular settings considered here.

As demonstrated, the ability of the developed agents to reproduce fundamental behaviors such as the speed-density relation as the emergent phenomena significantly simplifies the simulation process and expands the limits of encompassable exploratory scenarios. The developed model not only makes is easier to simulate the scenarios such as earthquake and inundation but also improve the accuracy by enabling the agents to best use the available space and include different degrees of debris distribution. In order to reproduce equivalent results with 1D models or simple agent-based models, a large number of observations are necessary to set the evacuees' speeds, according to the degree of resistance caused by the debris. Further, the above presented nighttime evacuation scenario, which demands a detailed model of the environment and sophisticated constituent functions to model the interactions in a wide open spaces with limited visibility, visitors, etc., cannot be simulated with simplified models, emphasizing the need of sophisticated simulators like the one being developed herein.

Pedestrian-car mixed mode evacuation with non-signalized junctions

As a demonstrative application of the junction trajectories presented in Section 3.3.2 and the constituent functions to make the agents interact at non-signalized junctions (i.e., $g^{car\_at\_non-signalized\_junction}$), we simulated hypothetical evacuation scenarios using the same Kochi city environment, restricting the agents to the streets (i.e., open spaces like parks, are excluded). The emergency evacuation is advised at 11:00 p.m., and 61,218 people are expected to evacuate to over 10 m elevation. The initial locations of the agents are set according to the building occupancy data of the simulated region. To quantitatively evaluate the advantages of allowing cars for evacuation, we set a given p population percentage using cars, assuming that each car carries three persons. The slowest agents at least 1 km away from the nearest evacuation area were assigned to cars. We consider the following five scenarios:

1. All the agents walk
2. slowest 6% are allowed to use cars; total of 1224 cars

3. slowest 9% are allowed to use cars; total of 1836 cars
4. slowest 15% are allowed to use cars; total of 3060 cars
5. 15% of the slowest were allowed to use cars, and some roads were restricted either only to pedestrians or cars to reduce the interactions between them

The statistics of the speeds and evacuation start times are assigned according to Table 4.7. Since this is a night-time evacuation, we assumed that people preferred to walk on roads wider than 4 m; the pedestrian agents are allowed to use roads of any width but with higher preference for roads wider than 4 m when available, according to Algorithm 3. We assumed that the traffic signals are inoperable due to a hypothetical condition. No earthquake damages or tsunami inundation is considered. We assume that the car users behave rationally and follow decent driving habits. Further, the car agents are assumed to give priority to pedestrians.

In all the five scenarios, pedestrian agents were prevented from entering the evacuation areas reserved for car agents to prevent congestion at the entrances, with the number of cars to each evacuation area controlled to reduce long traffic jams and to respect the parking space limitations. While in the first four scenarios, the agents could choose the closest evacuation area without any route choice restrictions, in the 5th scenario, some road stretches were restricted to either allow only cars or pedestrians to travel along those road stretches. However, at junctions, pedestrians (cars) were allowed to cross a road dedicated for cars (pedestrians). This restriction was expected to reduce the friction between cars and pedestrians making it possible to use large number of cars. Also, as in Table 4.7 in all the scenarios, car agents were forced to start evacuation within first 20 minutes of the time of tsunami warning, mimicking the restriction of evacuation start time window for the car users as a strategy to reduce friction between cars and pedestrians.

Figures 4.23 and 4.24 show some snapshots of the agents' movements at junctions. Figure 4.23 shows the merging movements of cars at a junction and the slowing of traffic at a sharp bend. Close examinations show the movements of cars along curved trajectories, avoiding collision with the cars on other trajectories. Figure 4.24 shows some snapshots of pedestrians affecting the flow of cars.

Figure 4.25 compares the evacuation progress. According to Fig. 4.25c, when no cars are allowed, about 2% of the population could not reach the target evacuation area within 70 minutes. This small percentage comprised the slow-moving people located over 1 km away from the nearest safe area. Conversely, allowing 6% of the population to use cars enabled 100% of the population to reach their destinations within 60 minutes. In addition, the 6% use of cars slightly accelerated the evacuation progress. Although the use of 9% and 15% cars also accelerated the evacuation progress during the first 40 minutes, thereafter, numerous pedestrians on the roads force the cars to move slower; see Fig. 4.25b. We observed that the pedestrians significantly

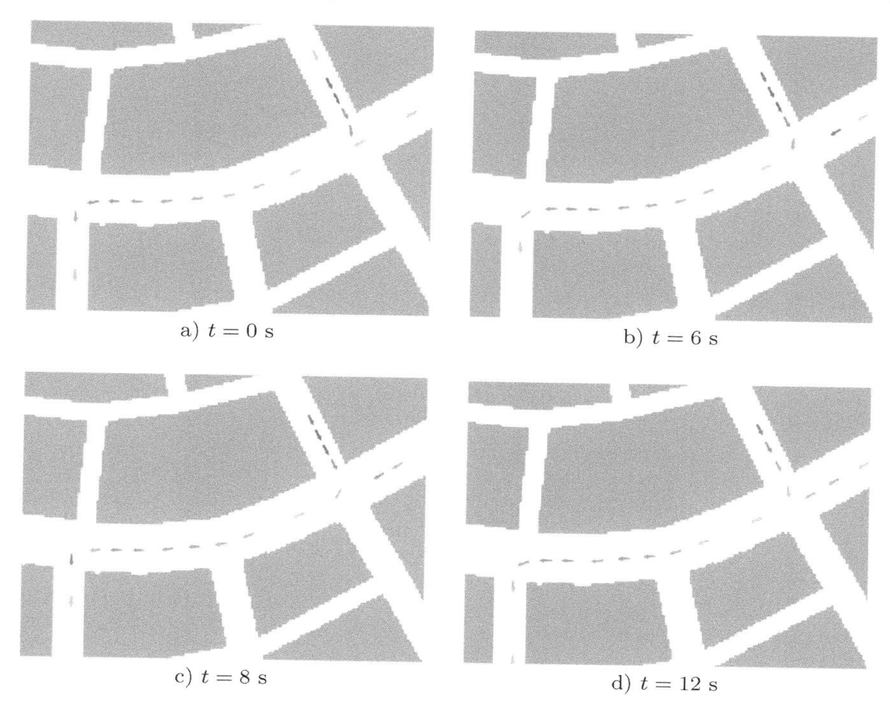

FIGURE 4.23: Snapshots of simulated traffic congestion at junctions.

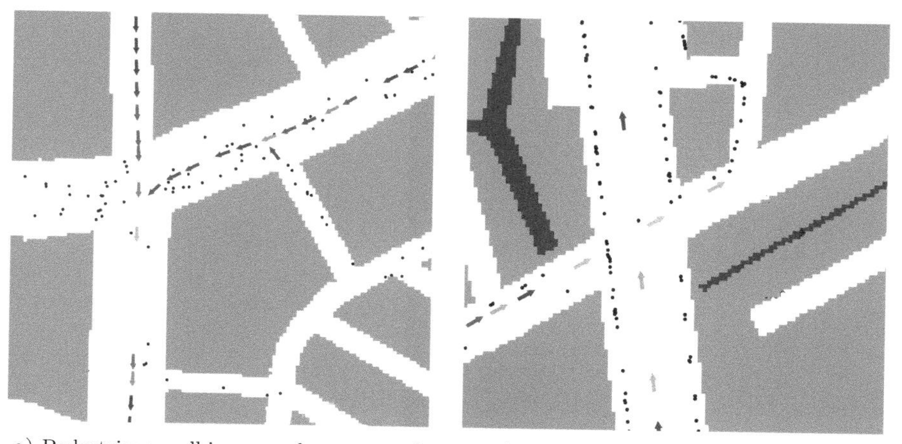

FIGURE 4.24: Traffic congestion caused by pedestrians at non-signalized junctions.

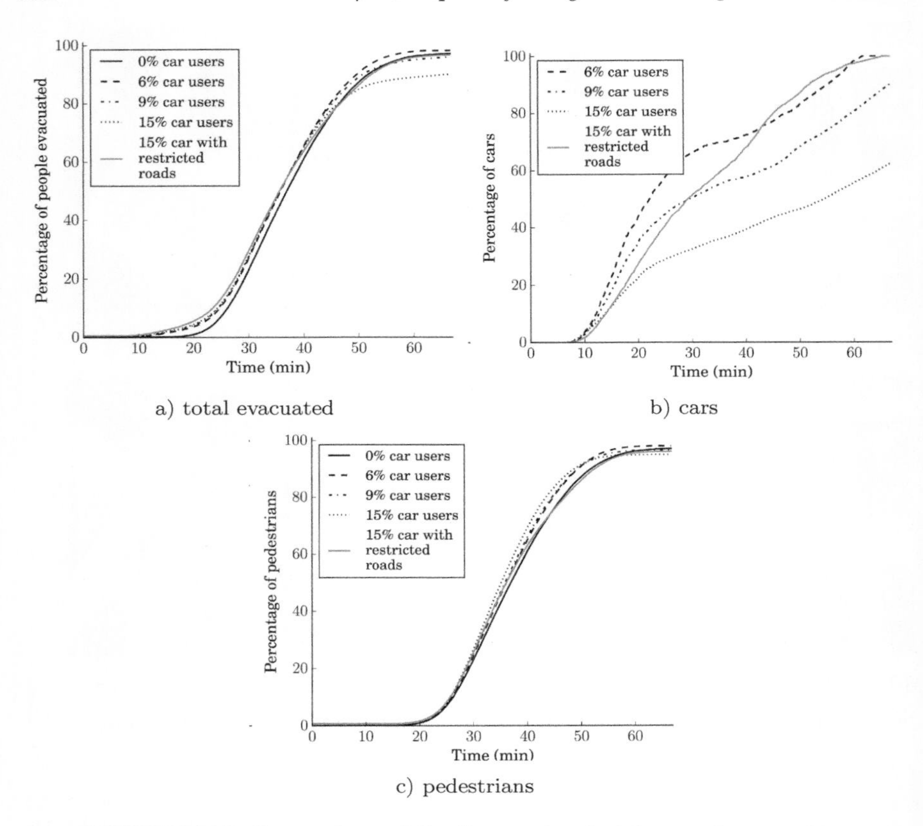

FIGURE 4.25: Comparison of the five car+pedestrian mode scenarios.

influenced the cars, especially at junctions along the routes shared by both pedestrians and cars; see Fig. 4.24. As seen in Figs. 4.25a and 4.25b, 15% cars significantly lowered the total number evacuated due to these congestions.

In the scenario 5, we tried to reduce this pedestrians' interruption to traffic flow by restricting the usage of some stretches of roads either to pedestrians or cars. This restriction is flexible, i.e., pedestrians (cars) are allowed to cross any of the restricted roads at junctions. Hence, enforced restrictions do not eliminate the car-pedestrian interactions. As seen in Fig. 4.25b, this restriction dramatically improves the evacuation progress making most of the evacuees to reach a shelter within 60 minutes. Though the progress is slower compared to 6% and 9%, its performance is comparable to that with no cars, indicating that many cars can be allowed if planned well. The road restrictions in the scenario 5 is set manually observing the level of pedestrian induced car delays in a series of simulations. Experts would be able to device better performing plans based on their experiences, or by automating the search for better performing road restrictions.

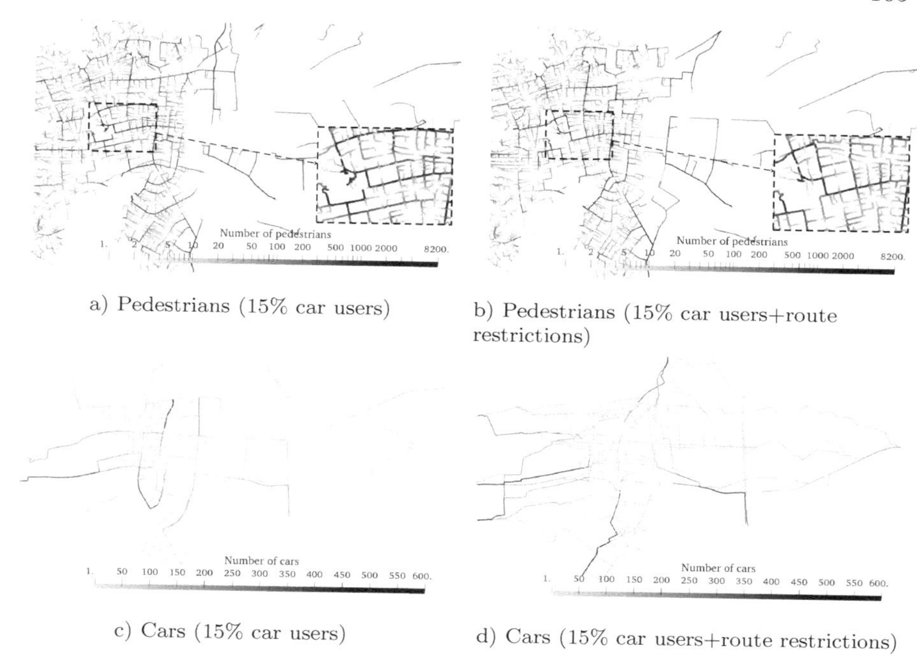

a) Pedestrians (15% car users)

b) Pedestrians (15% car users+route restrictions)

c) Cars (15% car users)

d) Cars (15% car users+route restrictions)

FIGURE 4.26: Routes taken by pedestrians and cars for the last two scenarios with 15% car users.

According to Figs. 4.26a and 4.26b, the changes in path usage by pedestrians in the scenarios 4 and 5 are minor and localized to the center of the city. Conversely, Figs. 4.26c and 4.26d indicate that the car routes have significantly changed, and some cars were forced to travel more than 3 km to reach a shelter. Figure 4.27 compares the statistics of each agent's characteristic speed and the average speed, estimated as the ratio of the distance traveled and time taken to reach its destination. In all the scenarios, a negligible difference between the characteristic and average pedestrian speeds are observed, possibly due to the high mobility of the pedestrian agents and the priority given to them at junctions. Unlike for the pedestrians, the estimated average speeds of cars are drastically different from their characteristic speeds. Though a noticeable difference was expected since cars must slow down at junctions even with empty roads, this drastic difference is due to the pedestrians' influence at junctions and along the roads. Figure 4.27h shows that the route restrictions in scenario 5 have significantly improved the progress of cars compared to the scenario 4, and the estimated average speed of cars in scenario 5 is almost close to that of scenario 2 with 6% car users.

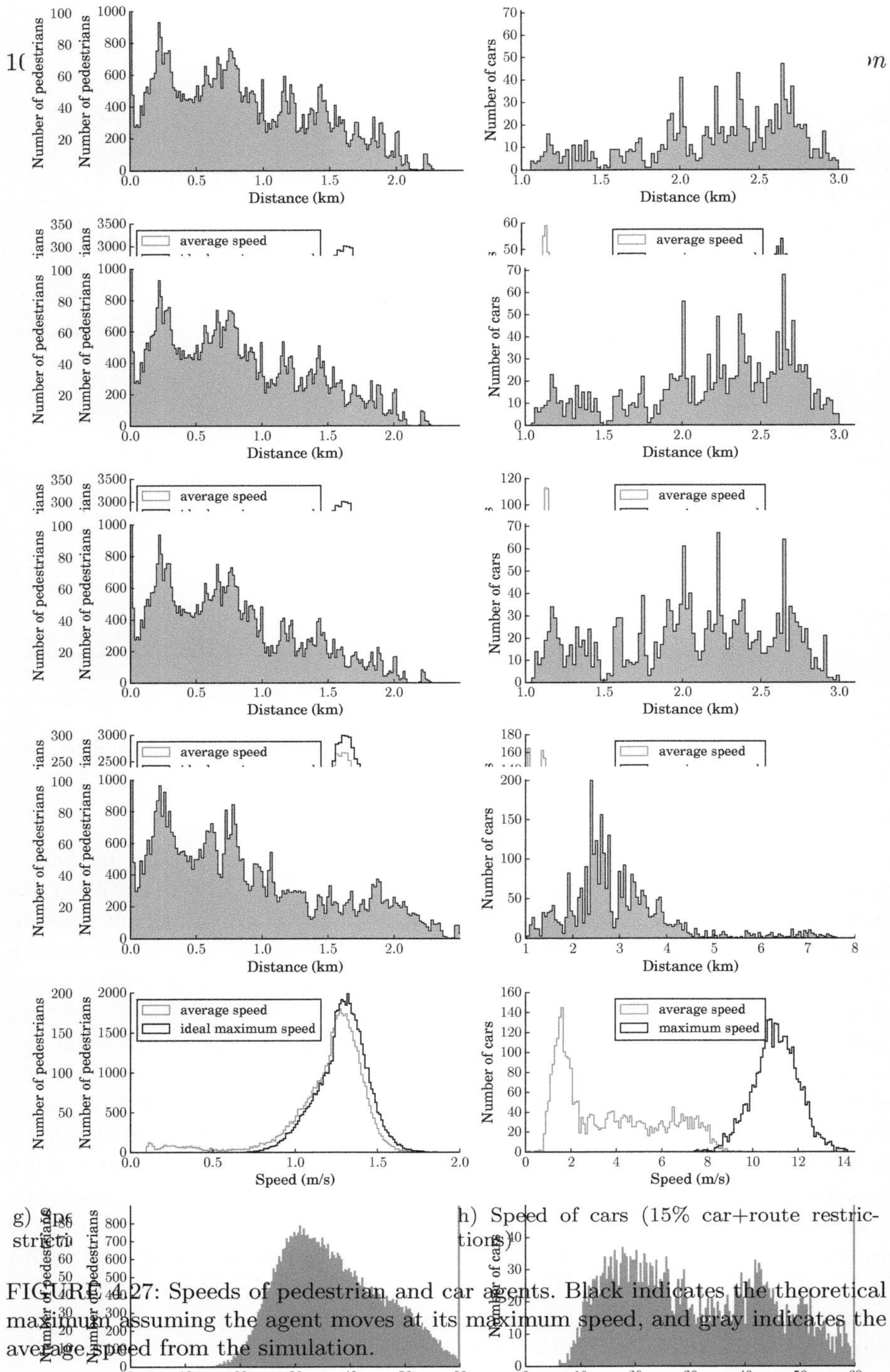

g) Spe... stricti...

h) Speed of cars (15% car+route restrictions)

FIGURE 4.27: Speeds of pedestrian and car agents. Black indicates the theoretical maximum assuming the agent moves at its maximum speed, and gray indicates the average speed from the simulation.

Remarks on the agent based evacuation simulations

As demonstrated, the agents' ability to produce fundamental relations, including speed-density, as an emergent behavior even for new exploratory scenarios, for which prior observations are unavailable or observations are hard to obtain, enables the evacuation planners to include all the influencing factors, including visibility, debris, complex conditions at junctions, etc., in comprehensively estimating certain evacuation scenario and their strategies to accelerate the evacuation process. While the list of available constituent functions, $g^{\cdots}$, must be expanded to encompass all the influencing factors, the constituent functions must be thoroughly tested even after minor improvements by comparing with the real observations or those from controlled experiments to ensure they can accurately reproduce the reality. Appendix B.3 provides details of parallel automated tool to find suitable values for the parameters of the constituent function to best reproduce a given observation (Leonel *et al.* 2014). Such automated parameter searching ability can significantly reduce the parameter tuning efforts, enabling the adoption of the agent-based system in regions with different social systems.

A

Conjugate Gradient Method

CONTENTS

DOI: 10.1201/9781003149798-A

In this appendix, the conjugate gradient method for the wave equation with preconditioner and parallel computation are explained, including the related basics, using the finite element method for earthquake simulation as an example.

A.1 Wave equation and its solution

To check if the discretization dependence of the numerical result is sufficiently small, the numerical solution is compared with the analytical and converged solutions that remain unchanged with the finer discretization. In such cases, the solution quality is typically examined in terms of numerical dispersion, unique to the wave equation (e.g., seismic wave propagation analysis).

Numerical dispersion is briefly explained by the 1D wave equation, $c^2\,\partial^2 u(x,t)/\partial x^2 = \partial^2 u(x,t)/\partial t^2$. Here, c and $u(x,t)$ are the wave speed and the displacement in space x and time t, respectively. The space domain of the 1D wave equation is discretized by an evenly spaced grid of ds, and $\partial^2 u(x,t)/\partial x^2$ is approximated by a second-order central difference, later considering the discretization effect. If the displacement is set to $\bar{u}\,e^{i\omega t - ikx}$, the numerical wave speed of the discretized equation is $\sin(\delta)/\delta\, c$, where δ is $k\,ds/2$. Compared with the wave speed c before discretization, c slows down with $\sin(\delta)/\delta$. When ds is small enough, the wave speeds before and after discretization are almost the same, but it slows down as ds increases. This indicates that the wave speed changes depending on the relationship between the wavelength and ds (i.e., numerical wave speeds with short wavelengths (in the high-frequency range where the accuracy is not guaranteed) are slower than those in the equation to be solved). The frequency-dependent dispersion due to such discretization is called numerical dispersion.

For waves with insufficiently discretized frequency components, the numerical simulation cannot be performed at the original wave speed due to numerical dispersion, thus warranting for the removal of such frequency components for quality assurance of the numerical simulation. The most common method involves converting the time waveform from the analysis into frequency components using the fast Fourier transform (FFT), removing the components not guaranteeing the accuracy, and then converting the rest back into the time domain using the inverse fast Fourier transform (high-cut filter). Considering slow or stagnant wave speeds of the undesired high-frequency components of the numerical simulation input, they must be removed by the high-cut filter before conducting the numerical simulation. Such filtering is used in linear cases, while in the nonlinear cases, the numerical solutions must be discretized at a finer level than the frequency component sufficient to evaluate the nonlinearity (e.g., constitutive law), and then the frequency components outside the accuracy guarantee must be handled based on the evaluation index.

As mentioned above, errors are used to evaluate the accuracy of the numerical results with the most common being the relative error. When comparing solutions of nt time steps, the relative error is calculated from the reference solution u_{it}^{ref} and the target solution u_{it}^{target} at it time steps as $\sqrt{\sum_{it=1}^{nt}(u_{it}^{ref} - u_{it}^{target})^2 / \sum_{it=1}^{nt}(u_{it}^{ref})^2}$ from u_{it}^{target} and u_{it}^{target} to evaluate the closeness of the reference and target solutions. However, even if one deviates from another by just one time step, such as $u_{it}^{target} = u_{it-1}^{ref}$, a large relative error may occur. Therefore, it is necessary to use different error definitions based on the application and purpose. E.g., misfit criteria focusing on the envelope shape and the waveform phase (e.g., Kristekova, Kristek and Moczo *et al.*, 2006) have been proposed.

A.2 Preconditioner

Consider solving the n-dimensional linear equation $\mathbf{Ax} = \mathbf{f}$ by the conjugate gradient method. Since the method uses conjugate vectors to expand the solution, it can reach the exact solution $\mathbf{x}$ in at most n iterations. Also, the conjugate gradient method chooses the conjugate vector closest to the steepest direction minimizing the error at that iteration. As a result, $\bar{\mathbf{x}}$ which satisfies criteria (e.g., relative error $\|(\mathbf{A}\bar{\mathbf{x}} - \mathbf{f})\|/\|\mathbf{f}\| < \epsilon$. ϵ is a positive small number (e.g., 10^{-8})) is expected to be reached in a fewer iterations than n (the history of how much the relative error is at how many iterations is called the convergence history, and the property (speed and stagnation of convergence) is called convergence).

Since the conjugate gradient method is generally used for a huge n (n reached 10^{9-12} in a recent finite element analysis), the convergence must be improved to obtain an approximate solution in a smaller number of iterations. To meet this expectation, the matrix equation to be solved is preconditioned for improving the convergence of the conjugate gradient process (preconditioner) so that the condition number (ratio of the largest eigenvalue to the smallest eigenvalue) is close to 1 and the width of the eigenvalue distribution is narrowed. Specifically, we construct the preconditioning matrix $\mathbf{M}$, expected to improve the performance, and process $\mathbf{Ax} = \mathbf{f}$ as $\mathbf{M}^{-1}\mathbf{Ax} = \mathbf{M}^{-1}\mathbf{f}$ to make $\mathbf{M}^{-1}\mathbf{A}$ the target of the conjugate gradient method, instead of $\mathbf{A}$ (left-handed preconditioning matrix. In some cases, there is also a right-handed preconditioning matrix applied from the right, such as $\mathbf{AM}^{-1}$). The above improvement in the condition number and eigenvalue distributions intuitively implies that $\mathbf{M}^{-1}\mathbf{A}$ is closer to the unit matrix, and $\mathbf{M}$ achieving this is the best-performing preconditioner. Obviously, using $\mathbf{A}$ as $\mathbf{M}$ is the best preproconditioner, but this is meaningless. In the end, the best performing $\mathbf{M}$ is the one close to $\mathbf{A}$ that has a small inverse matrix evaluation cost. Although many different $\mathbf{M}$ have been proposed, the choice of $\mathbf{M}$ depends on the tradeoff between the degree of improvement in convergence and the construction cost of $\mathbf{M}$ itself and the cost of applying it within the conjugate gradient process. As explained in the latter half of this section, Point and Block Jacobi methods are not so effective but with almost no construction and application costs, so we always apply preconditioner when solving the problem. In addition, although rare, if the condition number of $\mathbf{A}$ is extremely bad, it may be difficult to accurately calculate the conjugate vector, the premise of the conjugate gradient method, and convergence cannot be guaranteed (error reduction stops halfway and stagnates). In such cases, it may be possible to ensure convergence by applying a high-performing preconditioner. If convergence cannot be ensured even in such a case, the calculation precision may be insufficient (calculation usually done in double precision), and convergence may be achieved by changing the calculation precision to quadruple or more.

Herein, as the preconditioner examples, Point Jacobi and Block Jacobi methods, which are not so effective but with almost no construction and application costs, and the Incomplete LU and Incomplete Cholesky methods, highly effective but with high construction and application costs, will be explained. When solving dynamic problems in an earthquake field using the finite element method, one rough guide to determine the characteristics of $\mathbf{Ax} = \mathbf{f}$ is the mesh size, physical properties, and the size of time increments. Although not exact, if these are close to the Courant condition, the properties of $\mathbf{A}$ are not so bad, and if they are farther away, the properties often become worse. Specifically, the former corresponds to the analysis of seismic wave propagation or ground motion amplification, while the latter corresponds to the analysis of structures or soil-structure systems. We decide which preconditioner must be used after considering the total cost of analysis (cost of construction and application and number of iterations). For example, when Newmark's β method is used as the time integration, the mass matrix with $1/(\text{time step size})^2$ is included near the diagonal term of $\mathbf{A}$, and the convergence property is much better than the static problem. Conversely, in earthquake problems, the nonlinear time history response is often analyzed for 10^{3-5} time steps, wherein the preconditioning matrix needs to be reconstructed at every time step (or once every few time steps). In the finite element method, the computational cost per iteration tends to be relatively small compared to that of the preconditioning matrix because $\mathbf{A}$ is a sparse matrix. Therefore, for the large construction cost of the preconditioning matrix, the improvement of the convergence property by preconditioning is unworthy. In addition, many problems in the earthquake field (soil-structure system analysis) have extremely different effects on the convergence properties of $\mathbf{A}$. In such cases, a method with low construction and application costs and not so strong convergence performance improvement effect can be applied to the soil, and a method with high construction and application costs but strong convergence performance improvement effect can be applied to the structures. In general, this approach is feasible in almost every case, since the degree of freedom of the ground tends to be large as it covers a large area, while that of the structures is relatively small in many cases.

The Point Jacobi and Block Jacobi methods are often used in dynamic problems. The Point Jacobi method uses a diagonal matrix consisting of the diagonal terms of $\mathbf{A}$ as $\mathbf{M}$. Specifically, when the ij-elements of $\mathbf{M}$ are m_{ij} and the ij-elements of $\mathbf{A}$ are a_{ij}, set $m_{ii} = a_{ii}$ (diagonal terms) and the other terms (non-diagonal terms) to 0. In this case, $\mathbf{M}^{-1}$ can be easily calculated because the diagonal term is set to $1/a_{ii}$ and the non-diagonal terms are set to 0. The Block Jacobi method performs the Point Jacobi method on the diagonal Block matrix. The following is an example of a 3D finite element analysis wherein the displacement responses of the x_1, x_2, x_3 components are discretized in the x_1, x_2, x_3 coordinate system, commonly used in an earthquake field (no boundary condition processing is assumed for simplicity). In this case, $\mathbf{A}$ is a set of 3×3 submatrixes, each of which can be written down as $a_{3(i-1)+k,3(j-1)+l}$. Each component of this submatrix represents the effect of the x_k component of the i-th node on the x_l component of the j-th node, and they represent components strongly dependent on each other (there is a strong relationship between the x_1, x_2, and x_3 components of the displacement of each node). The Block Jacobi method improves the properties of such strongly connected components by grouping them together. Specifically, only the small matrix $a_{3(i-1)+k,3(i-1)+k}$ of 3×3 on the diagonal with particularly strong ties is used as $\mathbf{M}$, while the rest are set to zero. In this case, $\mathbf{M}^{-1}$ can be constructed by calculating the inverse of $a_{3(i-1)+k,3(i-1)+k}$ of each submatrix of 3×3. Therefore, it is possible to construct a preconditioning matrix that is stronger than the Point Jacobi method without increasing the construction and application costs. As mentioned above, since $\mathbf{A}$ contains a mass matrix with $1/(\text{time step size})^2$ near the diagonal term, some convergence performance improvement is expected. Therefore, it can be useful in many practical cases, not only in seismic wave propagation analysis and ground motion amplification analysis, but also in soil-structure analysis considering the total balance of construction and application costs: the total number of iterations to solve $\mathbf{Ax} = \mathbf{f}$ in one time step (the number of iterations per time step is expected to be much smaller than for the static problem), the computational cost per iteration ($\mathbf{A}$ is a sparse matrix). In this explanation, the submatrix is 3×3 because the unknowns per node are the x_1, x_2, x_3 displacement components, but if the number of unknowns per node is larger, the submatrix and the preconditioner can be constructed in the same way.

Incomplete LU and Incomplete Cholesky are preconditioners with high construction and application costs but may improve convergence performance and are used for the dynamic and static problems with extremely poor convergence. These are often used in finite element analysis generating symmetric matrices, and the conjugate gradient method with Incomplete Cholesky is called the Incomplete Cholesky Conjugate Gradient (ICCG) method. Incomplete Cholesky corresponds to the symmetric matrix case of Incomplete LU, thus Incomplete LU will be described next.

TABLE A.1: Discretization setting and efficacy of preconditioners.

(a) Discretization setting: node coordinate & connectivity

node #	1	2	3	4	5	6	7
x	0	5	9	1	7	3	10

element #	1	2	3	4	5	6
left-side node #	1	4	6	2	5	3
right-side node #	4	6	2	5	3	7

(b) Improvement of eigenvalue distribution and condition number (ratio of maximum eigenvalue to minimum eigenvalue) by each preconditioner in discretization settingD.

matrix	λ_1	λ_2	λ_3	λ_4	λ_5	condition #
$\mathbf{A}$	2.	1.81	1.31	0.691	0.191	10.5
$\mathbf{U}^{-1}\mathbf{L}^{-1}\mathbf{A}$	1.	1.	1.	1.	1.	1.
$\bar{\mathbf{U}}^{-1}\bar{\mathbf{L}}^{-1}\mathbf{A}$	1.43	1.	1.	1.	0.571	2.5
$Diagonal(\mathbf{A})^{-1}\mathbf{A}$	1.82	1.41	1.	0.591	0.184	9.90

Incomplete LU (ILU) uses an incomplete application of the lower-upper (LU) factorization as a preconditioner, so we first describe LU factorization and then explain its incomplete application. The following boundary value problem (e.g., a static bar) can be solved using the finite element method as:

$$\frac{d^2u(x)}{dx^2} = 0 \quad (0 < x < 10), \quad u(0) = 0, \quad u(10) = 1. \tag{A.1}$$

Discretize this equation using linear elements, set the nodes and connectivity as Table A.1-a, and perform boundary condition processing to obtain $\mathbf{Ax} = \mathbf{f}$ as the equation to be solved, where $\mathbf{x}$ is unknown variable on each node,

$$\mathbf{A} = \begin{pmatrix} 1 & 0 & 0 & -1/2 & -1/2 \\ 0 & 3/2 & 0 & -1/2 & 0 \\ 0 & 0 & 3/2 & 0 & -1/2 \\ -1/2 & -1/2 & 0 & 1 & 0 \\ -1/2 & 0 & -1/2 & 0 & 1 \end{pmatrix}, \quad \mathbf{f} = \begin{pmatrix} 0 \\ 1 \\ 0 \\ 0 \\ 0 \end{pmatrix}.$$

In the LU factorization, $\mathbf{A}$ is first decomposed into a product of lower triangular matrix, $\mathbf{L}$, and upper triangular matrix, $\mathbf{U}$, such that $\mathbf{A} = \mathbf{LU}$. In this example, we have the following:

$$\mathbf{L} = \begin{pmatrix} 1 & 0 & 0 & 0 & 0 \\ 0 & 1 & 0 & 0 & 0 \\ 0 & 0 & 1 & 0 & 0 \\ -1/2 & -1/3 & 0 & 1 & 0 \\ -1/2 & 0 & -1/3 & -3/7 & 1 \end{pmatrix},$$

$$\mathbf{U} = \begin{pmatrix} 1 & 0 & 0 & -1/2 & -1/2 \\ 0 & 3/2 & 0 & -1/2 & 0 \\ 0 & 0 & 3/2 & 0 & -1/2 \\ 0 & 0 & 0 & 7/12 & -1/4 \\ 0 & 0 & 0 & 0 & 10/21 \end{pmatrix}.$$

Next, using the decomposed $\mathbf{L}$ and $\mathbf{U}$, solve for $\mathbf{Ly} = \mathbf{f}$, and then solve for $\mathbf{Ux} = \mathbf{y}$ to obtain $\mathbf{x}$. In this case, since $\mathbf{L}$ is a lower triangular matrix, $\mathbf{Ly} = \mathbf{f}$ can be easily solved by sequentially substituting the solutions from the top. Also, since $\mathbf{U}$ is an upper triangular matrix, $\mathbf{Ux} = \mathbf{y}$ can be solved by sequentially substituting the solutions from the bottom. Thus, if we can divide into the form $\mathbf{A} = \mathbf{LU}$, we can easily apply the operation corresponding to $\mathbf{A}^{-1}$.

Comparing $\mathbf{A}$ with $\mathbf{L}$ and $\mathbf{U}$, some of the components zero in $\mathbf{A}$ now have values (e.g., 41 and 14 components). The pattern of non-zero components in the original matrix is called the non-zero pattern, and placing the values in the components not included in the non-zero pattern (components zero in the original matrix) is called fill-in. In a complex 3D problem, even if $\mathbf{A}$ is a sparse matrix, there will be vast fill-in in $\mathbf{L}$ and $\mathbf{U}$. Thus, to use the direct method for solving the matrix equation, we will need more computer memory to store the matrices in the solution process than expected, despite the original target matrix being sparse. Conversely, standard conjugate gradient finite element analysis codes take advantage of the matrix sparsity to store the non-zero component values and their positions in the matrix using a data compression format, such as Compressed Row Storage, used to realize the vector products associated with $\mathbf{A}$ in the conjugate gradient method. Here, we consider performing an incomplete LU decomposition taking advantage of the matrix sparsity. In other words, since we have information on the non-zero pattern of the sparsity of $\mathbf{A}$, we can use it to obtain an approximate $\bar{\mathbf{L}}$ and the upper triangular matrix, $\bar{\mathbf{U}}$ without allowing fill-in to the parts where $\mathbf{A}$ is zero (The degree to which fill-in is allowed is expressed by changing X in ILU(X). ILU(0) not allowing fill-in at all is described here as ILU). This is an approximate solution method, and while it does not give the exact solution using $\bar{\mathbf{L}}$ and $\bar{\mathbf{U}}$, it can ensure some constraint on the original solution space of $\mathbf{A}$. From a computational point of view, ILU(0) is an attractive choice, although there is some degradation in the preconditioning performance. This is because the resulting $\bar{\mathbf{L}}$ and $\bar{\mathbf{U}}$ are sparse matrices comparable to $\mathbf{A}$ requiring less application cost and computer memory and it avoids the need to allocate memory each time a fill-in occurs ad hoc during the construction of $\mathbf{L}$ and $\mathbf{U}$, and the pressure on computer memory caused by the large number of fill-ins. In the actual application, $\mathbf{M}^{-1} = \bar{\mathbf{U}}^{-1}\bar{\mathbf{L}}^{-1}$ is calculated as $\bar{\mathbf{U}}^{-1}$ and $\bar{\mathbf{L}}^{-1}$. The computation of $\bar{\mathbf{U}}^{-1}$ and $\bar{\mathbf{L}}^{-1}$ is similar to that of the LU decomposition above, so the application itself is straightforward (in serial computation).

Table A.1-b shows the efficacy of each preconditioning. Without preprocessing, $\mathbf{A}$ has the wide distribution of eigenvalues and large condition number. $\mathbf{U}^{-1}\mathbf{L}^{-1}\mathbf{A}$ applied with $\mathbf{U}^{-1}\mathbf{L}^{-1}$, equivalent to the inverse of $\mathbf{A}$, is naturally a unit matrix and shows the best performance. The properties of $\bar{\mathbf{U}}^{-1}\bar{\mathbf{L}}^{-1}\mathbf{A}$ applied with $\bar{\mathbf{U}}^{-1}\bar{\mathbf{L}}^{-1}$ preprocessed by ILU are much better than those of $\mathbf{A}$. Conversely, $Diagonal(\mathbf{A})^{-1}\mathbf{A}$ from the Point Jacobi method applied with $Diagonal(\mathbf{A})$, having only the diagonal component of $\mathbf{A}$, remains almost the same as $\mathbf{A}$. This tendency is because the target problem is a static problem and there is no mass matrix with 1/(time step size)^2 near the diagonal term, typically appearing in dynamic problems. Since the effect of preconditioning varies depending on the matrix characteristics, it is important to choose the right one depending on the problem characteristics.

A.3 Finite element with parallel computation

Here, finite element analysis using parallel computing is described. First, the recent computer environment for parallel computing is described, and then the method of performing parallel finite element analysis on such computer environment is explained.

The configuration of a recent computer used for numerical simulation generally consists of a CPU and computer memory accessed by the CPU (in some cases, accelerators such as GPUs are added). This single computer configuration is called a computer node, and a cluster computer consisting of 10^{1-4} computer nodes connected by a high-speed network are often used in recent parallel computing. In such a computing environment, computer memory is not shared among the computer nodes (distributed memory environment), thus it is necessary to operate the computer nodes as if distributed memories were shared. Specifically, computation results are synchronized among computer nodes as if they were computed on a single computer. In such a distributed memory environment, Message Passing Interface (MPI), a communication standard is often used for communication among computer nodes, and the parallel computation based on MPI is called MPI-based parallel computation. Since this communication needs to be explicitly incorporated into the code, and in some cases, the data structure in the code needs to be rewritten for MPI parallel computation, often acquiring huge cost for this implementation. In contrast, computer nodes often contain multiple CPU sockets and multiple CPU cores in a single CPU socket, and parallel computation within a computer node using these sockets is also possible. The CPU cores on multiple CPU sockets can access the computer memory in the computer node, and the implementation of intra-node parallelization is easy because it does not require communication as in MPI parallelization. Open Multi-Processing (OpenMP) is a typical method for implementing intra-node parallelization. With OpenMP, parallel computation can be implemented by simply inserting directives into the existing code with the advantages of requiring little code modification, no loss of code portability or visibility, and low cost of implementing parallel computation. Such parallel computing using OpenMP is called OpenMP parallel computing. Although OpenMP parallel computing is easy to implement, the maximum number of parallel computations is limited to the number of CPU sockets $\times$ CPU cores in a computer node, and the available computer memory is limited to the amount that can be installed in a computer node. In addition, in OpenMP parallelization, when multiple CPU cores access the shared computer memory in parallel, delays may occur due to conflicts, possibly degrading the parallel computation performance. In such cases, it is necessary to increase the independence of access to the memory (e.g., coloring). In actual parallel computing, OpenMP/MPI hybrid parallel computing is often used, where OpenMP is used for intra-node parallel computing and MPI is used for inter-node parallel computing.

There are two measures of parallel computing performance, strong scalability and weak scalability, and which measure is more critical depends on the application. Strong scalability indicates how fast the analysis time is when the same problem is solved in different parallels. The weak scalability indicates how the analysis time changes when the number of parallels is increased without changing the problem size per parallel unit. For example, in strong scalability, the computation time is plotted for a single problem to be analyzed and the number of nodes, $256 \times 2^{i-1}$, is changed, to see how the computation time is improved by the number of parallels. In weak scalability, the size of the problem is determined for 256 computer nodes, and the computation time is checked to see if the computation time changes when 2^{i-1} times the size of the problem is solved on $256 \times 2^{i-1}$ computer nodes. The main causes of scalability degradation are communication for MPI parallel computation, memory access conflict during OpenMP computation, differences in the granularity of each parallel computation, and serial computation parts that are difficult to parallelize.

Finite element analysis using the conjugate gradient method with MPI parallel computing is explained using a simple example. Since the introduction of OpenMP parallelization can be achieved by inserting typical directives into a typical FEM code, and the issues arising when introducing OpenMP parallelization to some non-typical cases (e.g., memory access conflict related to element-by-element method) is too advanced, we will focus only on MPI-based parallel computation.

First, we construct an ordinary finite element model used without parallel computation. For clarity, a simple model with 14 elements and 14 nodes is used here (Fig. A.1-a). It is common to generate the elements using a single

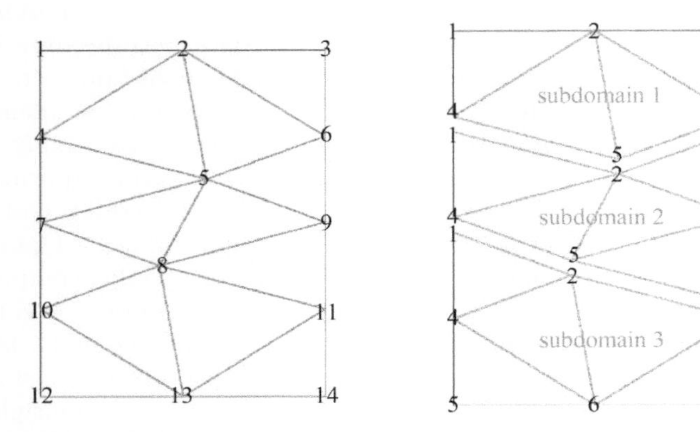

a) original finite element model b) decomposed finite element model

FIGURE A.1: Finite element model with parallel computation.

computer node with large computer memory (in some cases, using OpenMP parallel). The original finite element model is divided into the number of parallel computer nodes to generate a parallel finite element model, and the conjugate gradient method for parallel computation (Algorithm 3) is used to obtain the solution. Here, the number of computer nodes is set to three and the finite element model is divided into three subdomains (Fig. A.1-b): subdomain 1 (4 elements, 6 nodes), subdomain 2 (5 elements, 7 nodes), and subdomain 3 (5 elements, 7 nodes). In performing this partitioning, the computational load of the subdomains must be equal (uniform granularity) and the number of nodes (nodes means nodes of finite element model) shared among the subdomains must be reduced, as shown in the following example.

In the following, we will use Algorithm 3 and Fig. A.1 to explain this in detail. The objective is to show how the partitioned finite element model in Fig. A.1-b can be solved in a distributed memory environment as if the model in Fig. A.1-a were being solved. First of all, Algorithm 3 is composed of the basic operations of numerical computation: matrix-vector product, vector inner product, and vector addition and scalar multiplication. The vector addition and scalar multiplication do not need to be synchronized among computer nodes (can be calculated within a subdomain). The vector inner products, α_i and β_i, are the calculation results in the entire domain, thus all computer nodes need to communicate with each other. E.g., in the subdomain j, $(\mathbf{r}_i^j, \mathbf{p}_i^j)$ is calculated, and then it is communicated to all nodes and added together to calculate $(\mathbf{r}_i, \mathbf{p}_i)$. This type of communication is called global communication with a large communication load. Although the amount of communication for a single inner product of this Algorithm 3 is small, it can significantly degrade the parallel performance when a large number of computer nodes are used because all of them communicate.

The matrix-vector product is described next. To realize the calculation of $\mathbf{Ax}_0$, equivalent to the case without parallel calculation, the results obtained by $\mathbf{A}^j\mathbf{x}_0^j$ in each subdomain are synchronized. Specifically, a communication table for inter-node communication is created, and information on nodal variables is communicated to the neighbors to synchronize the analysis performed in each subdomain. This local communication is called adjacent communication. In this case, information must be communicated to the nodes on the contact surface of each subdomain, and although this is a local communication between the neighboring domains, the amount of communication is large, thus the communication must be controlled as in the inner product calculation using global communication. The details are explained below using Fig. A.1-b as an example. Subdomain 1 is adjacent to subdomain 2, and the information to be sent from subdomain 1 to 2 is the nodal variables of nodes 4, 5, and 6 of subdomain 1. This is then sent to subdomain 2 and added to nodes 1, 2, and 3 of subdomain 2. Subdomain 2 is adjacent to subdomains 1 and 3, and the information to be sent from subdomain 2 to 1 is the nodal variables of nodes 1, 2, and 3 of subdomain 2. This is then sent to subdomain 1 and added to nodes 4, 5, and 6 of subdomain 1. The information to be sent from subdomain 2 to 3 is the nodal variables of nodes 4, 5, and 6 of subdomain 2. This is then sent to subdomain 3 and added to nodes 1, 2, and 3 of subdomain 3. Subdomain 3 is adjacent to subdomain 2, and the information to be sent from subdomain 3 to 2 is the nodal variables of nodes 1, 2, and 3 of subdomain 3. This is then sent to subdomain 2 and added to nodes 4, 5, and 6 of subdomain 2.

This may seem complicated first, but becomes easier to understand when it is organized as a process in each subdomain, so we will emphasize this and organize it as follows. First, before starting the analysis, the node numbers sending nodal variables from subdomain j to subdomain i are listed with the node number of subdomain j ($sendnode_i^j$). It also lists the node number in subdomain j to which the nodal variables sent from subdomain i to subdomain j are added by the node number of subdomain j ($merge_i^j$). Within the conjugate gradient process, soon after computing $\mathbf{A}^j\mathbf{x}_0^j$ in the subdomain j, the nodal variable information is packed using $sendnode_i^j$ and sent to the subdomain i. On the other hand, in subdomain j, unpack the nodal variables received from subdomain i and add them to the corresponding nodal variables by $merge_i^j$. By synchronizing each subdomain in this way, $\mathbf{A}^j\mathbf{x}_0^j$, calculated independently in each subdomain, it may appear as if $\mathbf{Ax}_0$ is being calculated. In practice, the implementation is a bit more contracted, but can be written in a simple way as shown above.

From the above, the first primary factor determining the performance of MPI parallel computation for finite element analysis is the uniformity of granularity (the amount of computation of $\mathbf{A}^j\mathbf{x}_0^j$ and vector addition and scalar multiplication). If the granularity becomes non-uniform, the processing time of computer nodes becomes non-uniform. As a result, the parallel performance is degraded because it is synchronized with the slowest processing time. Second is the amount of communication between the computer nodes. Although nodal

Algorithm 3 Conjugate gradient method in subdomain j with parallel computation to find the numerical solution of the matrix equation $\mathbf{Ax} = \mathbf{f}$ with relative error $(\|\mathbf{r}_i\|/\|\mathbf{f}\| < \epsilon)$. ^{j} indicates variables in subdomain j.

$\mathbf{t}^j \Leftarrow \mathbf{A}^j \mathbf{x}_0^j$ (MatVec: local communication)
$\mathbf{r}_0^j \Leftarrow \mathbf{f}^j - \mathbf{t}^j$
$\mathbf{p}_0^j \Leftarrow \mathbf{r}_0^j$
$i \Leftarrow 0$
while $(\|\mathbf{r}_i\|/\|\mathbf{f}\| \geq \epsilon$ **do**
  $\mathbf{t}^j \Leftarrow \mathbf{A}^j \mathbf{p}_i^j$ (MatVec: local communication)
  $\alpha_i \Leftarrow \frac{(\mathbf{r}_i, \mathbf{p}_i)}{(\mathbf{p}_i, \mathbf{t})}$ (DotProduct: global communication)
  $\mathbf{x}_{i+1}^j \Leftarrow \mathbf{x}_i^j + \alpha_i \mathbf{p}_i^j$
  $\mathbf{r}_{i+1}^j \Leftarrow \mathbf{r}_i^j - \alpha_i \mathbf{t}^j$
  $\beta_i \Leftarrow \frac{(\mathbf{r}_{i+1}, \mathbf{t})}{(\mathbf{p}_i, \mathbf{t})}$ (DotProduct: global communication)
  $\mathbf{p}_{i+1}^j \Leftarrow \mathbf{r}_{i+1}^j - \beta_i \mathbf{p}_i^j$
  $i \Leftarrow i + 1$
end while

information only on the contact surface is exchanged between the subdomains, a vast data needs to be sent and received, degrading the parallel computing performance, so attention must be paid to how to reduce the number of nodes for communication in finite element model.

As described above, when the number of degrees of freedom that can be computed on a single computation node is n, the introduction of MPI parallel computing is expected to lead to a significant improvement in analysis capability, since problems with approximately $m \times n$ degrees of freedom can be solved using m computer nodes. For earthquake simulations, the introduction of MPI parallel computing is beneficial because of numerous degrees of freedom required to perform the analysis on a fine mesh, where the mesh dependence of the numerical solution is sufficiently small to guarantee the quality of the simulation. Conversely, the introduction of MPI computation as described above may require large code modification costs due to changes related to data structure, communication, etc., and the parallel computation performance may deteriorate for computations with degrees of freedom where the communication cost is relatively large compared to the computation cost of $\mathbf{A}^j$ for each subdomain. Thus, the advantages and disadvantages of introducing MPI must be carefully considered. For example, when capacity computing to solve various medium-scale problems for uncertainty quantification is performed using m computer nodes, if this medium-scale problem can be computed on a single node with the recent increase of computer memory and CPU cores per node, it may be better to use OpenMP intra-node parallelization instead of MPI parallelization since the former requires less installation cost and computation time.

B

Multi-Agent System

CONTENTS

DOI: 10.1201/9781003149798-B

To increase the reliability of the ABM, every factor significantly influencing the evacuation progress must be adequately modeled so that corresponding field observations can be reproduced to a sufficient accuracy. The formation of crowds, speed variation according to crowd density, and the slowing down of vehicles when sharing the road space with the pedestrians are some examples of such phenomenon with high impact on evacuation time. As mentioned in Section 2.3, our objective is to tune the developed ABM to reproduce these phenomena as emergent behaviors of individual agent's actions and interactions, allowing detailed explorations of the scenarios. This appendix presents the details of the implemented collision avoidance model, different agent interaction models based on the collision avoidance model, and tuning of the parameters of interaction models to reproduce given field observations.

B.1 Collision avoidance

Different collision avoidance schemes have been proposed by several researchers, such as the social force model (Helbing *et al.* 1995) and other force-based models (Cristiani *et al.* 2014), quad-tree based path planning approaches (Ghoshray *et al.* 1996), probabilistic maps (Isler *et al.* 2005), Optimal Reciprocal Collision Avoidance (Berg *et al.* 2011) among others. Some of these collision avoidance schemes are not created specifically to solve the collision avoidance problem but to navigate avoiding collisions (as a by-product for efficient navigation) or specifically introduce the implicit effects among the entities avoiding collisions (social force model). Out of these schemes, the Optimal Reciprocal Collision Avoidance (ORCA) (Berg *et al.* 2011) was implemented in our ABM.

This section presents an overview of the collision avoidance scheme and its implementation, and exemplifies the problems of adopting the ORCA scheme in the ABM along with the strategies to overcome them. The contents discussed in this appendix are from Dr. Leonel Aguilar's PhD. thesis, "Enhancements and applications of a scalable multi-agent based large urban area evacuation simulator with emphasis on the use of cars", submitted to the Dept. of Civil Engineering, the University of Tokyo.

B.1.1 Brief introduction to ORCA scheme

For conciseness, only a brief explanation of ORCA scheme is presented here, and interested readers are referred to the original work by Jur van den Berg *et al.* (2011) for details. The physical extent of an agent a is modeled as a disc with a given radius r_a; though any convex shape (e.g., ellipse) can be used, circular shaped is preferred due to the simplicity in calculations. Even a partial overlap of two disks can make the two agents collide. The objective of a

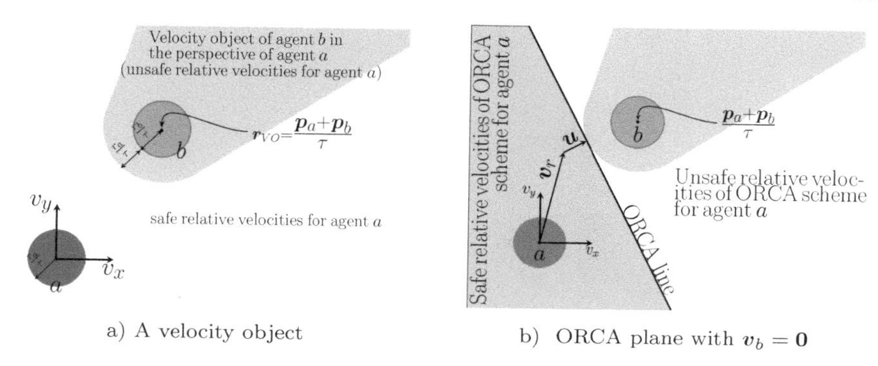

a) A velocity object

b) ORCA plane with $\boldsymbol{v}_b = \mathbf{0}$

FIGURE B.1: ORCA scheme for two agents a and b. $\boldsymbol{p}_i$, r_i, and $\boldsymbol{v}_i$ are the position, radius and velocity of agent $i \in \{a, b\}$.

collision avoidance algorithm is to find suitable speeds and moving directions for agents so that they will not collide within a desired period of time τ, while making minimum deviations from their desired moving directions and speeds. The desired moving direction and speed can be defined as those taken by the agent when no agents in the vicinity are obstructing its movements.

ORCA algorithm first identifies the set of all relative velocities that bring a given agent a into a collision state with another agent b. For a pair of agents, all the possible collision velocities form a convex shape with one side extending to infinity, referred to as a *Velocity Object* (VO). Figure B.1a shows the velocity object of an agent b from agent a's perspective. The space is scaled by $\frac{1}{\tau}$ since we desire the agents to move collision-free for a period of τ. Assuming the agent b continues moving at its current velocity $\boldsymbol{v}_b$, ORCA algorithm finds the smallest change $\boldsymbol{u}$ in the relative velocity $\boldsymbol{v}_r$ that would bring the two disks into contact. The point at which the smallest change $\boldsymbol{u}$ touches the velocity object $\boldsymbol{V}_c = \boldsymbol{p}_a + \boldsymbol{v}_r + \boldsymbol{u}$, and $\boldsymbol{V}_c$ is the closest point to VO from $\boldsymbol{p}_a + \boldsymbol{v}_r$ (see Fig. B.1b), where $\boldsymbol{p}_a$ and $\boldsymbol{p}_b$ are the positions of the two agents. In the original ORCA scheme, each agent is made to take complementary action to avoid collision by making one agent to adjust the velocity by $\alpha\boldsymbol{u}$ and the other by $(1-\alpha)\,\boldsymbol{u}$. $\alpha = 0.5$ makes each agent behave reciprocally by making each agent to put an equal effort to avoid collision.

The smooth convex velocity object with simple geometric properties, formed by a single agent, makes it straight forward to find the optimal change in velocity $\boldsymbol{u}$ (i.e., least effort to avoid collision). Conversely, Fig. B.2 shows that the union of the velocity objects formed by multiple neighboring agents is neither convex nor geometrically simple, making it difficult to find the optimal solution. When multiple velocity objects are involved, the ORCA algorithm opts for a feasible solution with low computational cost instead of finding the optimal $\boldsymbol{u}$ from the non-convex solution space of the original problem. Hence, contrary to the name, the ORCA scheme is not guaranteed to find the optimal solution when more than one neighboring agents are present.

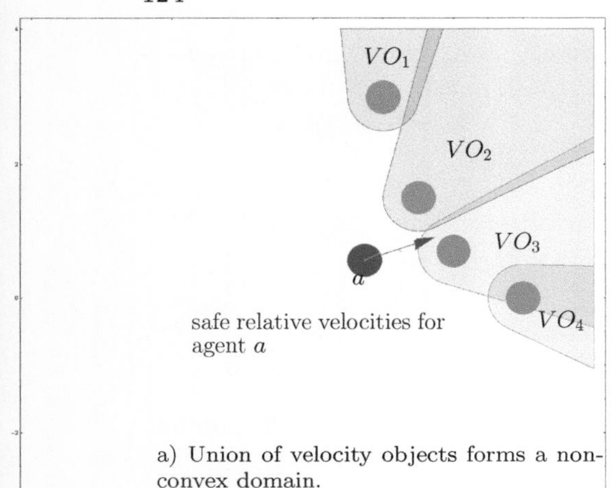

a) Union of velocity objects forms a non-convex domain.

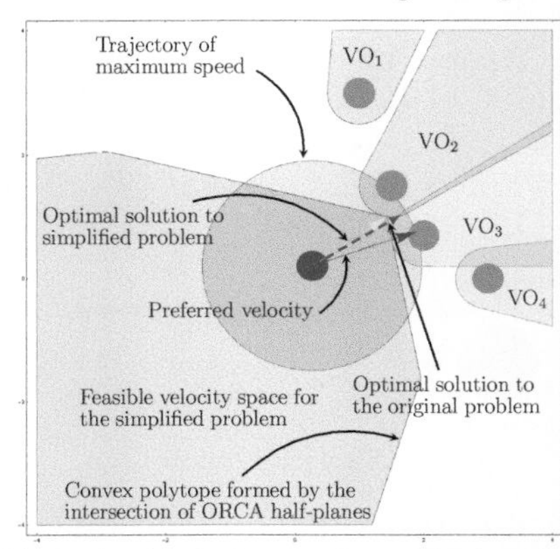

b) Optimal solution to the original problem, and the optimal solution to the simplified problem posed by ORCA scheme.

FIGURE B.2: When many neighboring agents are present, ORCA scheme sacrifices the optimality for computational efficiency by posing a simplified problem.

To find a near optimal solution with less computational effort, the ORCA scheme first simplifies the shape of the safe velocity space into a convex polytope, and find the optimal solution to the simplified problem using linear programming. In posing an easily solvable problem, ORCA scheme first defines a convex sub-space of the safe-velocity space of the original problem as the intersection of half-planes. First a tangent to a velocity object is drawn at $\boldsymbol{V}_c$. This tangent line is referred to as an ORCA-line, and the half-space excluding the velocity object (i.e., the half-plane containing only collision-free velocities) is defined as the feasible half-plane. Figure B.1 shows the ORCA line and the corresponding feasible half-plane with only one neighboring agent. When many neighboring agents are present, a convex polytope is defined by taking the intersection of the half-planes from each of the velocity objects, as show in Fig. B.2b. This convex polytope defined the space of the feasible velocities to the simplified problem (i.e., all possible collision-free velocities to the simplified problem). The ORCA scheme finds the optimal solution to the newly posed simplified problem using linear programming.

In a multiple-agent scenario, the shapes of the velocity objects are characterized by the geometries of the two entities avoiding collisions and the parameter τ. The quantities involved in the ORCA line generation are the radius of current agent r_a, the radius of current neighbor r_b, the expected collision-free travel time τ, and the amount of adjustment α. In the two-agent case, the obtained velocity can be assured to be collision-free only if the following three conditions are satisfied: both agents follow the same collision avoidance scheme; each agent takes the complementary responsibility percentage α of each other; the assumption of a constant relative velocity $\boldsymbol{v}_r$ is maintained during τ time.

As seen in Fig. B.2b, the solution space of the simplified problem can exclude a significant portion of the original solution space, and often this excluded portion containing the optimal solution to the original problem. Hence, the optimal solution to the simplified problem might not be the optimal of the original problem. Irrespective of this exclusion of a portion of the original solution space, ORCA algorithm often finds a near optimal solution (for example, see Fig. B.2b). Also, the small time increment Δt (e.g., we used Δt =0.2 s) and ensuring the collision avoidance at each time increment contribute to the ORCA scheme's success of producing near optimal solutions. In sections B.1.4 and B.1.5, we explains two situations which can produce poor solutions and how to handle those.

While the parameter α makes the agents bear shared responsibilities to avoid collision, it makes agents depend on each other. Since our objective is to make the agents fully autonomous, we set $\alpha = 1$, making every agent take full responsibility of collision avoidance.

B.1.2 Implementation

In our implementation, we introduced several changes to the ORCA scheme to interface with the developed agents and the environment model. Although, the authors of the ORCA algorithm propose a way to incorporate the environment in the collision avoidance scheme (Berg *et al.* 2011), it isn't suitable for the hybrid environment used in our ABM (see Section 2.3). The modifications we introduced to interface the ORCA scheme with our ABM are briefly explained in the rest of this sub-section.

Agents must avoid collision not only with fellow agents but also with the obstacles in the environment. While the obstacles in the environment can be incorporated into the ORCA scheme by defining a suitable set of ORCA-planes, avoidance of collision with agents and environment can be treated in two ways. If the collision avoidance with agents and environment are treated as two independent problems, several iterative calls to respective collision avoidance functions are required to obtain a near optimal change in velocity, $\boldsymbol{u}$, to avoid collision with both the obstacles and agents. If it is treated as one problem by including ORCA half-planes from both agents and obstacles, the additional half-planes from obstacles can further reduce the extent of the solution space, leading to poor solutions. Thought the first is computationally expensive, it produces better solutions since its solution space less restricted compared to the second. Irrespective of that, in our implementation, we treated the collision avoidance with agents and obstacles as a monolithic problem, since it requires lesser computational effort. For avoiding the collision with obstacles, a set of half-planes, referred to as *navigational half-planes*, are created and used in the ORCA scheme just as the ORCA half-planes of agents.

As explained in Section 2.3.3, an agent scans the grid and identifies the boundary of its visibility (see Fig. 2.6). Analyzing the boundary of visibility

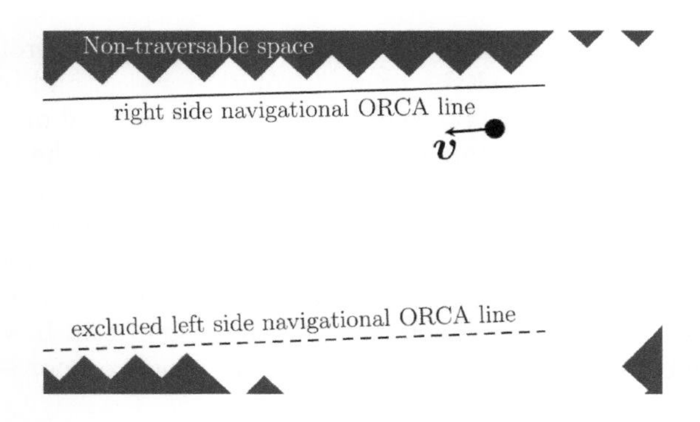

FIGURE B.3: Navigational half-planes for avoiding obstacles.

and comparing with the graph data, an agent can identify the boundaries of its path, and the boundaries of any obstacles on its way. If any of these boundaries are located within the $\tau \times v$ distance from the agent, a navigational half-plane is created for that boundary, where v is the preferred moving speed of the agent. As an example, when moving along a road, two navigational planes are generated as shown in Fig. B.3. In this example, no navigational half-plane in the left side of the agent is generated since the left side road boundary is farther than $\tau \times v$.

Under some combinations of the locations and movements of the neighboring agents, the intersection of the feasible half-planes can produce extremely small or empty feasible velocity space, making the agents move at very low speeds or bringing them to a complete halt. To avoid such undesired behaviors, the ORCA half-planes are ordered as per the distance from the corresponding neighbor. When the ORCA scheme produces a poor solution, a better solution is repeatedly sought by eliminating the half-plane with the lowest priority. In setting the priority order of half-planes, the navigational half-planes are given the highest priority, and the priority of half-planes of the neighboring agents are set inversely proportional to the distance. The half-planes leading to physically impossible results like agents moving over others must not be excluded. Moreover, only the agents within a certain proximity are considered in collision avoidance, since human counterpart also does not react to all the visible neighbors. This interaction with agents in a certain proximity reduces the computational cost.

Pedestrians and cars have physical limitations in acceleration and change in direction, and these limitations may depend on the age category, terrain, etc. To respect the limit of turning angle, we include two half-planes, called *steering half-planes*, at an angle of $\pm\Delta\theta$ to the agent's current moving direction, where $\Delta\theta$ is the maximum allowable change in direction within Δt period. If the current moving direction of the agent is $\boldsymbol{v}$, the two steering ORCA lines have the direction $\frac{1}{|\boldsymbol{v}|}\left(\boldsymbol{v} \pm \tan\left(\Delta\theta\right) \boldsymbol{k} \times \boldsymbol{v}\right)$, where $\boldsymbol{k}$ is the unit normal to the plane of the 2D environment and $\times$ stands for the cross product. To avoid any

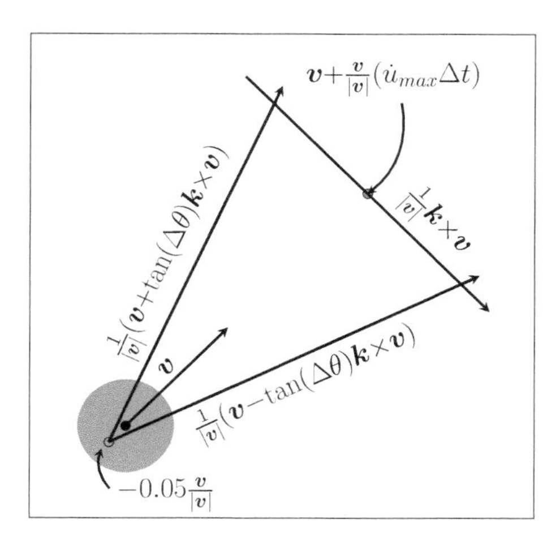

FIGURE B.4: Steering half-planes for respecting the physical constraints in rotation and acceleration.

problems caused by limited floating point precision, these two lines are made to pass through a point just behind the agent as shown in Fig. B.4. As an example, we set the steering ORCA lines to pass through $-0.05\frac{\boldsymbol{v}}{|\boldsymbol{v}|}$. To respect the limits in maximum acceleration, an ORCA line with the direction $\frac{1}{|\boldsymbol{v}|}\boldsymbol{k}\times\boldsymbol{v}$ is inserted through the point $\boldsymbol{v}+\frac{\boldsymbol{v}}{|\boldsymbol{v}|}(\dot{v}_{max}\Delta t)$, where $\dot{v}_{max}$ is the maximum possible acceleration.

Additional parameters are introduced to fine tune the usage of the collision avoidance: a boolean *overtake* to introduce preferences, such as if the evacuee desire to overtake another; their preferred side of overtaking, *side*, which is set to be the right side for all the cars; and the minimum distances an agent prefers to maintain between other agents, d_{pref}. Examples on using these parameters to implement specific interaction models are presented in Section B.2. Some combinations of parameters and situations can produce ORCA half-planes which contradict with the agents' physical constrains (maximum acceleration, maximum speed, etc.). To prevent such contradictions, every ORCA half-plane undergoes a simple feasibility test and the parameters are relaxed when the feasibility test is not satisfied.

B.1.3 Defining velocity objects and ORCA half-planes

This section provides a brief overview of the mathematical description velocity objects and ORCA half-planes required for computer implementation. An interested reader is referred to the original paper (Berg *et al.* 2011) for the details. The VO's perimeter consists of three regions; a circular arc segment, and two line segments tangent to the each ends of the arc. Let the agent avoiding the collision be a, the agent to be avoided be b, and their respective

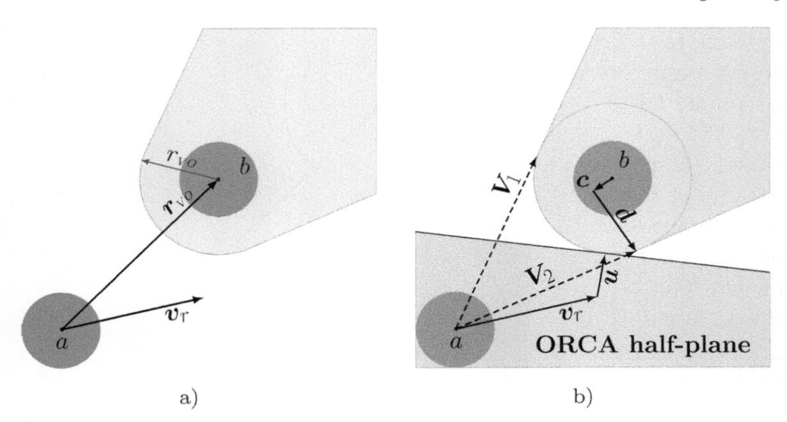

FIGURE B.5: ORCA half-planes creation. Note: ORCA half plane in (b) is with $\boldsymbol{v}_b = \mathbf{0}$.

velocities, positions and radius be $\boldsymbol{v}_a$, $\boldsymbol{v}_b$, $\boldsymbol{p}_a$, $\boldsymbol{p}_b$, and r_a, r_b. Further, let the ORCA parameters be α and τ.

$$\boldsymbol{r}_{VO} = \frac{1}{\tau}\left(\boldsymbol{p}_b - \boldsymbol{p}_a\right)$$

represents the relative vector from the center of a's physical space to the center of the circular region of the VO. $\boldsymbol{r}_{VO}$ can be interpreted as the relative velocity required by a to move to the location of b with in a period of τ seconds. The radius of the circular region of the VO is $r_{VO} = \frac{1}{\tau}\left(r_a + r_b\right)$. These two parameters, $\{\boldsymbol{r}_{VO}, r_{VO}\}$, completely define the VO's circular arc. We can identify the two points where the two straight boundaries of VO touch this circle using basic planar geometry. The lengths of the two vectors $\boldsymbol{c}$ and $\boldsymbol{d}$ in Fig. B.5 can be evaluated as $|\boldsymbol{c}| = \frac{r_{VO}^2}{|\boldsymbol{r}_{VO}|}$ and $|\boldsymbol{d}| = \sqrt{r_{VO}^2 - \boldsymbol{c}\cdot\boldsymbol{c}} = \frac{r_{VO}}{|\boldsymbol{r}_{VO}|}\sqrt{\boldsymbol{r}_{VO}\cdot\boldsymbol{r}_{VO} - r_{VO}^2}$, respectively. Let $\boldsymbol{r}_{VO}^u = \frac{1}{|\boldsymbol{r}_{VO}|}\boldsymbol{r}_{VO}$ and $\boldsymbol{r}_{VO}^n$ be the unit normal to $\boldsymbol{r}_{VO}^u$ (i.e., $\boldsymbol{r}_{VO}^u.\boldsymbol{r}_{VO}^n = 0$). Then, the two points $\boldsymbol{V}_1$and $\boldsymbol{V}_2$ at which the straight edge tangentially touches the circular boundary can be defined as

$$\begin{aligned}\boldsymbol{V}_{1,2} &= \left(|\boldsymbol{r}_{VO}| - \frac{r_{VO}^2}{|\boldsymbol{r}_{VO}|}\right)\boldsymbol{r}_{VO}^u \pm \boldsymbol{r}_{VO}^n\frac{r_{VO}}{|\boldsymbol{r}_{VO}|}\sqrt{\boldsymbol{r}_{VO}\cdot\boldsymbol{r}_{VO} - r_{VO}^2}\\ &= \left(\frac{r_{VO}}{\boldsymbol{r}_{VO}\cdot\boldsymbol{r}_{VO}}\right)\left(\left(\frac{\boldsymbol{r}_{VO}\cdot\boldsymbol{r}_{VO}}{r_{VO}^2} - 1\right) \pm \sqrt{\boldsymbol{r}_{VO}\cdot\boldsymbol{r}_{VO} - r_{VO}^2}\boldsymbol{k}\times\right)\boldsymbol{r}_{VO}.\end{aligned}$$

The smallest change $\boldsymbol{u}$ in the a's relative velocity that would bring it to the critical collision state with b (i.e., the two disks touching each other) is the shortest normal vector to the VO's boundary from the tip of $\boldsymbol{v}_r = \boldsymbol{v}_a - \boldsymbol{v}_b$, the relative velocity of a with respect to b.

To define the ORCA line, which defines the feasible half-plane, we require the closest point on VO from the tip of $\boldsymbol{v}_r$ and the tangent vector to VO at that point. Let $\boldsymbol{V}_c$ and $\boldsymbol{V}_t$ be the closest point on VO and the tangent vector to VO at $\boldsymbol{V}_t$. The Algorithm 4 outlines how to find these two vectors. Additional parameters are introduced in this step to better control the agents. The parameter *overpass* specifies if an additional perturbation to $\boldsymbol{v}_a$ must be applied to make agent a overtake b, the parameter *side* defines the direction in which this perturbation must be applied and γ determines the perturbation magnitude. The values of these parameters are determined according to the interaction model to be reproduced.

Algorithm 4 Setting closest point to a velocity object and ORCA line direction, $\boldsymbol{V}_c$ and $\boldsymbol{V}_t$.

if (*overtake*) //is the agent a overtaking the agent b
    $\boldsymbol{v}_a = \boldsymbol{v}_a + side \cdot \gamma\,(\boldsymbol{k} \times \boldsymbol{v}_a)$

$\boldsymbol{v}_r = \boldsymbol{v}_a - \boldsymbol{v}_b$
$\boldsymbol{P}_1 = \frac{\boldsymbol{V}_1 \cdot \boldsymbol{v}_r}{\boldsymbol{V}_1 \cdot \boldsymbol{V}_1}\boldsymbol{V}_1$ //projection of $\boldsymbol{v}_r$ on $\boldsymbol{V}_1$
$\boldsymbol{P}_2 = \frac{\boldsymbol{V}_2 \cdot \boldsymbol{v}_r}{\boldsymbol{V}_2 \cdot \boldsymbol{V}_2}\boldsymbol{V}_2$ //projection of $\boldsymbol{v}_r$ on $\boldsymbol{V}_2$

if $|\boldsymbol{P}_1| < |\boldsymbol{V}_1|$ **and** $|\boldsymbol{P}_2| < |\boldsymbol{V}_2|$ **then**//closest point on the curved edge of VO
    $\boldsymbol{g} = \boldsymbol{v}_r - \boldsymbol{r}_{VO}$
    $\boldsymbol{V}_c = \boldsymbol{r}_{VO} + \frac{r_{VO}}{|\boldsymbol{g}|}\boldsymbol{g}$
    $\boldsymbol{V}_t = -\boldsymbol{k} \times \boldsymbol{g}$
else if $|\boldsymbol{P}_1| < |\boldsymbol{P}_2|$ **then** //closest point on the line segment 1 VO
    $\boldsymbol{V}_c = \boldsymbol{P}_1$
    $\boldsymbol{V}_t = \boldsymbol{P}_1$
else //closest point is on the line segment 2 VO
    $\boldsymbol{V}_c = \boldsymbol{P}_2$
    $\boldsymbol{V}_t = \boldsymbol{P}_2$

The smallest change in $\boldsymbol{v}_r$ to bring a in to contact with b is given by:

$$\boldsymbol{u} = \boldsymbol{V}_c - \boldsymbol{v}_r.$$

The feasible half-plane is defined by the ORCA line passing through the point $\boldsymbol{v}_a + \alpha\boldsymbol{u}$ and has the direction $\boldsymbol{V}_t$. As mentioned above, we set $\alpha = 1$ to make agent a take necessary actions to avoid collision with b without depending on any complementary action from b to avoid the potential collision.

B.1.4 Group collision avoidance

As explained above, the simplification of solution space by introducing feasible half-planes can lead to poor solutions. While the simplification of the solution space into a convex polytope is the strength of the ORCA scheme, it can cause severe problems. When simplifying the safe relative velocity space to a half-plane, it is ensured that the per agent optimal solution is included. However,

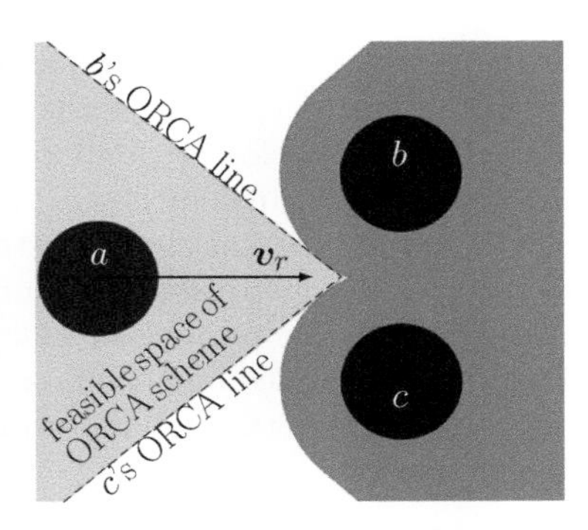

FIGURE B.6: Agent a chooses the optimal velocity from the safe velocities shown in light gray, making it impossible to pass the two stationary agents b and c. $\tau = 1$ s, $\alpha = 1$, $\boldsymbol{p}_a$ =(0, 0), $\boldsymbol{p}_b$ =(5, 1.8) m, $\boldsymbol{p}_c$ =(5, −1.8) m, $\boldsymbol{v}_a$ =(3.5, 0) ms^{-1}, $\boldsymbol{v}_b = \boldsymbol{v}_c$=(0, 0) ms^{-1}.

the union of local optima does not guarantee the inclusion of the global optimum collision-free velocity in the resulting convex polytope. In most cases, the ORCA scheme's solution is close to the optimal solution of the original problem (see Fig. B.2b). However, some combinations of neighbor agents' locations and velocities can lead to undesired behaviors like bringing the agent to a complete halt, forcing it to follow neighboring agents at slower speed, etc. To illustrate this, consider two stationary agents, b and c, positioned as shown in Fig. B.6, and an agent a wants to pass them avoiding any collision. Figure B.6, shows the resulting ORCA lines and the feasible velocity regions of ORCA scheme for agent a, choosing the optimal velocity from this region at each time step will inevitably lead it to get trapped behind agents b and c because of no space between them for a to pass through.

The best solution to this problem is to solve the original non-convex optimization problem, instead of searching for the optimal solution in ORCA scheme's simplified solution space. However, solving the resulting concave problem is complex and time-consuming. A numerically efficient alternative is to identify possible settings which can lead to poor solution during the construction of ORCA half-planes, and modify the half-planes to prevent the potential poor solutions.

To prevent the above explained problem of the agents behind a group of slow-moving agents, we introduced group collision avoidance. Closely spaced group of agents with no space for another agent to move, are identified and ORCA half-planes are created considering this group as a single entity, as shown in Fig. B.7d. As a simplification, three linear constraints form the group velocity object by identifying the farthest right, farthest left, and closest agent. This, although not optimal, allows the exclusion of undesired solutions leading evacuees to get trapped behind other evacuees, as seen in Fig. B.7e.

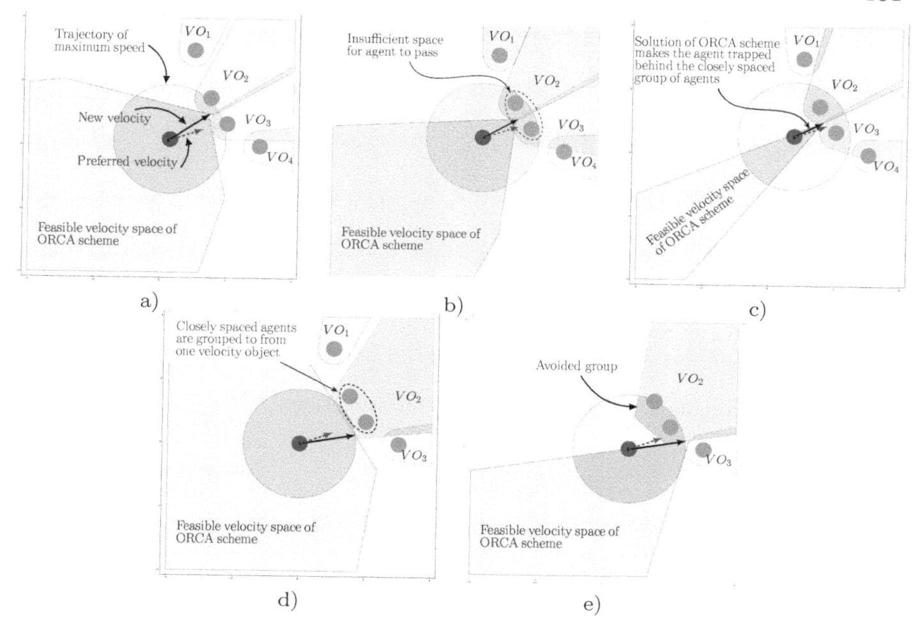

FIGURE B.7: Detection and generation of group velocity objects to prevent trapping of agents.

B.1.5 Side selection for overtaking

To provide an additional control over the creation of the ORCA half-planes, a mechanism giving priority to a certain side by adding a perturbation is introduced, as mentioned in Section B.1.3. This finer control is especially useful in avoiding situation such as getting trapped in the environment, see Fig. B.8, which is a similar condition to the problem described in Section B.1.4 but involving both obstacles and agents. To prevent trapping between obstacles and agents, it is checked whether there is a sufficient space between the agent to be overtaken and the obstacle. If there is no sufficient space in between, the availability of sufficient overtaking space on the other side of the agent is checked. If there is sufficient space on the other side, a small perturbation is added to the relative velocity to create the half-plane of the agent to be overtaken on the other side.

B.2 Interaction models

Collision avoidance alone is insufficient to model the pedestrians' and car-drivers' behaviors in an ABM. We introduced further parameters to have a

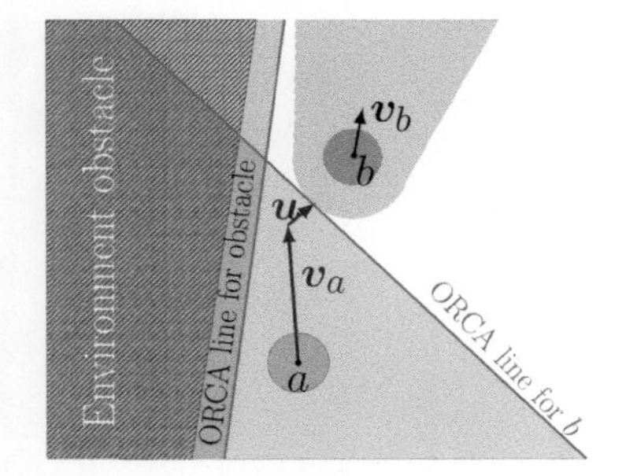

a) Space is insufficient to overtake b from left side

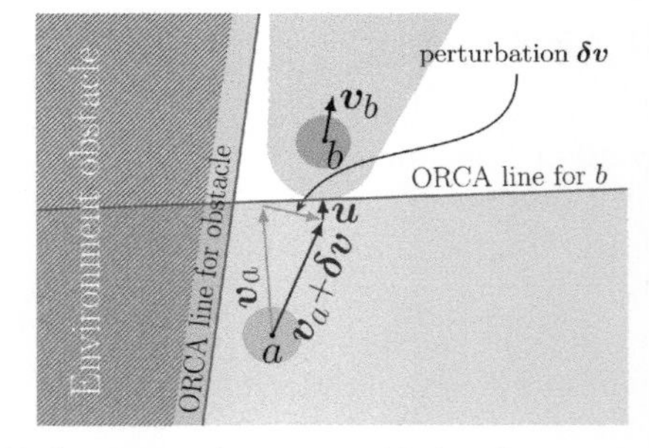

b) Small pertabation is added to favor a velocity toward right hand side

FIGURE B.8: Small perturbation, $\boldsymbol{\delta v}$, is added when space is insufficient to overtake.

finer control over the way agents take decisions on how to avoid collisions under different conditions. Although the rules and decision models (how the agent chooses its parameters) presented in this section are simple, they bring the simulation results closer to real-world observations. This section introduces the way these parameters are chosen or manipulated to represent specific behaviors.

B.2.1 Pedestrians

Pedestrian agents are modeled to interact asymmetrically with the neighboring agents on their back and front, as it is assumed that their awareness to actively avoid collisions is defined based on what they can see and consider as close, while their awareness of the agents on their back is based on the sensing of other agents through sound or other senses. Sound and other senses are not specifically included in our ABM. Instead, the back side agent at a close proximity is identified and interactions are modeled using this asymmetry. A real-world pedestrian does not respond to a highly sparse group and a dense group of pedestrians in the same manner. As an example, a pedestrian changes its direction much earlier when overtaking another in a sparsely distributed group of pedestrians compared to that in a denser group. To model such pedestrians' preferences according to the surrounding density, pedestrians in tight crowds will consider different collision avoidance parameters than those in the sparse situations. Specifically, we introduced a parameter to represent the preferred distance from the neighboring agents, and adjusted by τ according to the pedestrian density. These parameters are set according to the field observations, as explained in the above. Further, special care was given to the interactions between cars and pedestrians, as pedestrians actively try to get out of the cars trajectory if they sense danger.

B.2.2 Cars

Just like the pedestrians, the cars are modeled to avoid collision with anything lying in a certain visible proximity, which is set according to the density of the surrounding agents and at most equal to the agent's sight distance. τ is set according to the surrounding pedestrian density and speed in this case as well. To constrain a vehicle's motion to its current lane, half-planes are created to demarcate the traversable region of the lane. Steering half-planes restrict the maximum steering angle of vehicles and maximum acceleration (see Fig. B.4).

Car trajectories at junctions are approximated suing B-splines, as explained in Section 3.3.2, when traveling through a junction, navigational half-planes are included to make the cars move on the B-spline trajectory. Half-planes are inserted to make the car agents respect the maximum free flow speed within a junction, defined in Section 2.3.4. Further details on the car agents' collision avoidance at non-signalized junctions are explained in the above.

B.2.3 Calibrating the parameters to model specific interactions

Parameters and constants of interaction models must be set such that the agents can reproduce observed behaviors of the real-world counterparts. A specific set of parameters and constants to reproduce a given behavior and their respective algorithms represents a specific interaction model, and the process of finding the specific parameters is referred to as tuning or calibration. As a result of these interaction models, relationships between the agents interacting within a certain visible proximity and their speeds emerge. Following sections present the different specific interaction models and the emergent relationships contrasted with the observed data.

The interaction models presented in this section are calibrated with observations that are not made during evacuation situations. Also, we do not aim to provide a specific instance of the parameters but to demonstrate that the agents can be tuned to reproduce real-world observations through the suitable parameters. In the following three sections, we present calibration of agent parameters to reproduce observed fundamental diagrams (i.e., speed vs. density characteristics) for pedestrian-pedestrian interactions, car-car interactions, and car-pedestrian interactions. Section B.3 provides some details of the automated calibration tool used to search for the parameters.

Using the field observations reported by Mori *et al.* (1987), and with a regression over various observations compiled by Weidmann (Weidmann *et al.* 1993), we tuned the parameters of pedestrian agents for pedestrian-pedestrian interaction and validated comparing the results of the simulations with the respective field observations. The observation data from Mori *et al.* presents observations of unidirectional flows observed through the bird's eye camera in Osaka in walkways ranging from 2.2 m to 4.5 m in width and 20 m in length,

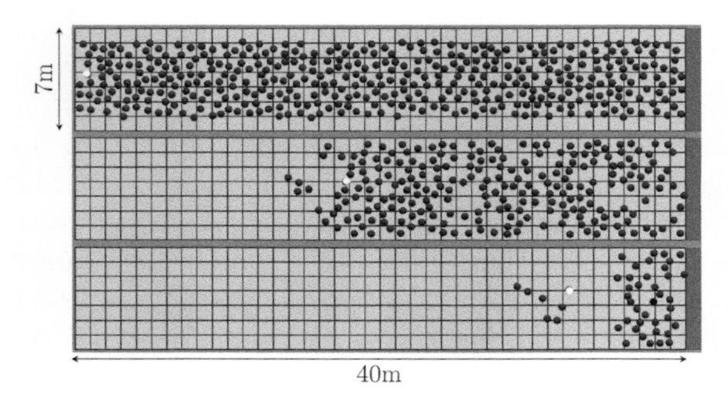

FIGURE B.9: Three snapshots of the scenario for validating pedestrian-pedestrian collision avoidance. Shown in white is an target agent whose speed and surrounding density are measured.

and an open boundary condition, while Weidmann's data as a major compilation of pedestrian research aggregates data from 25 research publications with different boundary conditions, different flow directions, and varying widths.

The validation scenario consists of a 5 m road with an additional 1 m on each side to simulate the effect of the open boundaries and 40 m length. A target agent is positioned among a crowd with a given maximum speed. Then, this agent and its surrounding crowd are set to evacuate in a unidirectional flow and its average speed is recorded, see Fig. B.9.

Using this setting, 25 Monte Carlo Simulations (MCS) are performed, one for each starting crowd density. Each MCS consisted in 100 simulations each. The average speed of the target agent is recorded for each simulation and its distribution is presented as a whisker-box plot per MCS along with a polynomial regression performed to data by Mori *et al.*, see Fig. B.10. The simulation results are in good agreement with the observations, especially up to the density 1.0.

Using the same validation scenario and re-tuning the parameters, the simulations results are compared with the observations reported by Weidmann (Weidmann *et al.* 1993). Figure 2.7a compares the numerical results with the reported observations. It is impossible to compare the dispersion of the current data with that of the data produced by the simulations, as only the regression over the data points is provided. The average behavior produces a good match with the regression reported by Weidmann. Additional fitting could be performed by introducing finer tuning capability in the parameter choosing phase by making further subdivisions in what the agent considers different crowd sizes, or by introducing a sense of available space in relationship by the number of surrounding agents instead as the plain number of surrounding agents.

The tuning and validation of car-car collision avoidance is performed by using a set of observation in Lincoln tunnel provided by Dhingara *et al.* (2008). Movement of car agents along a single lane of 1 km length is considered as the

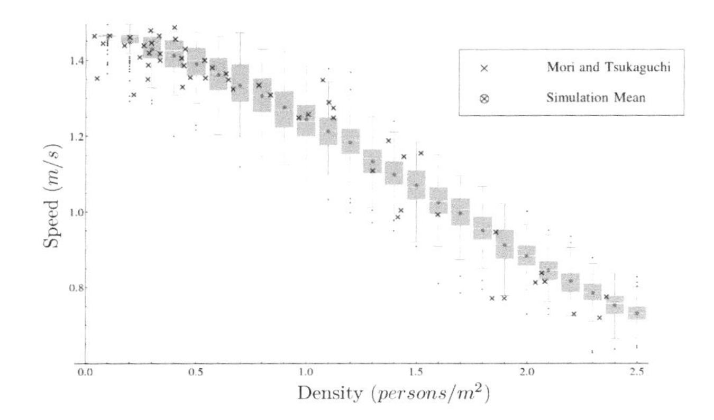

FIGURE B.10: Comparison of speed vs. density characteristics from the simulation with those from the field observations by Mori and Tsukaguchi (Mori *et al.* 1987).

problem setting. 100 simulations with different initial densities are performed. Samples of matching lane densities and current speed are obtained every 0.2 s, and the average over 20 s is reported as a data point. As seen in Fig. B.11, the simulation results are distributed around the observation data. While most points lay within 0.5 m/s distance from the observation data, some of the simulation results have 2 m/s disagreement with the observations. These outlying points can be further reduced by tuning using a larger set of observation data.

Most of the Japanese cities are full of narrow roads on which pedestrians and vehicles share space. Even on ordinary days, this is a major source of slow movements of vehicles. During a mass evacuation, such car-pedestrian interaction on narrow roads can be a major source hindering the car users. Though walking is the recommended mode of evacuation, numerous car users observed during the 2011 Great East Japan Earthquake and Tsunami made the disaster mitigation agencies to investigate the use of cars for evacuation, especially for the needy, such as elderly, mothers with children, etc.

Field observations

To quantify the effects of car-pedestrian interactions on narrow roads, we included a model of car-pedestrian interaction. Lack of real-world observations to quantify this interaction was a major difficulty we encountered. We conducted a small field observation at two locations on the narrow access road to Nezu shrine, Tokyo, Japan, during the golden week of 2013 to compile a data set for calibrating our interaction model. Many visit Nezu shrine during the golden week to enjoy the flowers. Videos of 72 cars passing through the surrounding crowds are recorded with a video camera (Sony Handycam HDR-CX170, 1080i). The aims of these observations are:

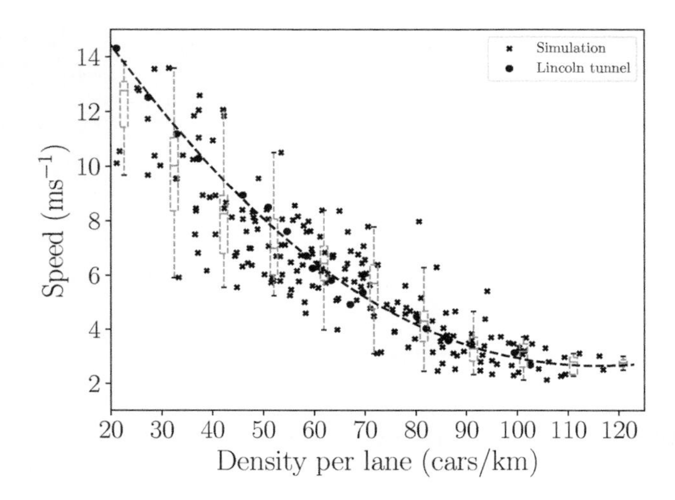

FIGURE B.11: Car-Car interaction validation results.

- Observe common behaviors of pedestrians and cars when interacting in a proximity to decide the rules for the car-pedestrian interaction model
- Find a quantitative relationship between pedestrians and cars for the validation of car-pedestrian interactions

The videos were recorded mounting the camera on a 3 m tall pole, at 2 locations. Fig. B.12a shows a frame of the videos taken from the second floor of a private house next to the street, which we refer as observation point 1. The second batch of observations were taken by the edge of the road, called the observation point 2, holding the pole on the pavement (see Fig. B.12b). Four square shapes of size 1 m×1 m were made on the road surface (see Fig. B.12a) so that the distance on the video frames can be accurately measured considering the perspective distortions. Two reference points marked the start and the end of the analysis area. The distance between these reference points is 12.25 m for observation point 1 and 19.7 or 18.22 for observation point 2 (depending on the visibility, as these marks were sometimes occluded from the street pole view).

The video recordings were trimmed to create a series of small videos of each car's movement between the start and the end of analysis area. Automatic trajectory tracking techniques could not be used since the available software could not handle video recordings at an oblique angle to the road surface. To perform the manual data analysis, two sets with different granularities are created. The first set is created with frames at 1 s interval (total of 1,496 frames) and the second set is created extracting frames at 0.1 s intervals (14,757 frames). From the 0.1 s frame batch, for each car the starting and the

a) Video frame from point 1

b) Video frame from point 2

FIGURE B.12: Sample video frames extracted from the observations. The rectangular markings on the road, seen in (a), are used to measure the distance considering the perspective distortion.

end frame are recorded. In the coarse grained frame batch for each car, the number of pedestrians in the vicinity Ω_t are counted.

The two quantities C^v and S^v are defined to quantify the video recording observations from the vehicles' perspective. C^v characterizes the state of the interacting neighborhood of a vehicle during its movement through the observation area, while S^v characterizes the average vehicle speed while moving through the observation area. As the physical environment remained constant throughout the observations, it does not appear in the construction of C^v.

Using the trimmed videos, the travel time of each car to move through the observation area was estimated. The average speed S^v is estimated by dividing the distance between the start and end points of the study area by the estimated travel time observation length.

$$S^v = \frac{\text{observation length}}{\text{observation time}} \approx \text{average speed.}$$

We define Ω_t as the region in which the presence of pedestrian will influence the behavior of a vehicle. For counting the number of pedestrian in Ω_t, we define the function $f_t(\boldsymbol{p})$ with binary output as

$$f_t(\boldsymbol{p}) = \begin{cases} 1 & \text{if } \boldsymbol{p} \in \Omega_t \\ 0 & \text{if } \boldsymbol{p} \notin \Omega_t, \end{cases}$$

where $\boldsymbol{p}$ represents the location of a pedestrian. Since no prior studies were available to decide the shape and the extent of Ω_t, we defined Ω_t based on our judgment of the recorded vehicle behaviors. Analyzing the video observations, it was estimated that the cars started slowing down with pedestrians in front of the vehicle within 3 m or less. Also, it was identified that the cars never leaving the road are delimited by the white lines. Based on these observations, the interaction region, Ω_t, is defined as a rectangle with a length equal to the lane width, and a width equal to 3 m (see Fig. B.13). The number of pedestrians in video frames at Δt interval was manually counted and stored as the vector

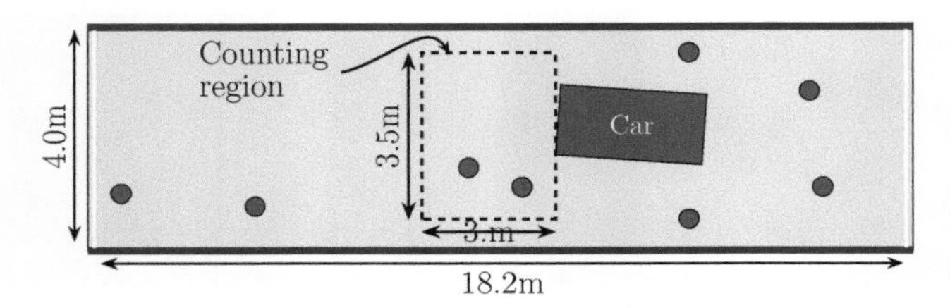

FIGURE B.13: Car counting.

$$F_t = \sum_{i=1}^{n} f_t(\boldsymbol{p}_t^i).$$

We define the cumulative pedestrians seconds, C^v, in the interaction area as

$$\begin{aligned} C^v &= \sum_{t=1}^{\frac{T}{\Delta t}} \sum_{i=1}^{n} f_t(p_t^i)\Delta t \\ &= \sum_{t=1}^{\frac{T}{\Delta t}} F_t \Delta t. \end{aligned}$$

C^v was estimated using $\Delta t = 0.2\,\mathrm{s}$. C^v for each car was aggregated into a graph without visible differences between the points coming from the different observation regions, implying that the differences in the two regions were not significant enough to be distinguishable with the amount and quality of the data collected (see Fig. B.14). We used these observations to tune the parameters of the car agents to model the car-pedestrian interactions.

Various dimensions of Ω_t were evaluated and the one presented in this section showed the best results. However, it is possible that better defined shapes would improve the robustness of the estimator. Given that $f_t \in \{0, 1\}$, $0 \leq F_t \leq n$, which guarantees that $\lim_{\Delta t \to 0} C^v$ converges. This highlights the robustness of the measure imposed on the state of the neighborhood, C^v.

Validation of car-pedestrian interaction

To validate the tuned model, we simulated above observation settings of a unidirectional flow of car agents in a bi-directional pedestrian flow. 100 simulations with different initial densities are performed, C^v is evaluated with a $\Delta t = 0.2\,\mathrm{s}$ over the period required for the vehicle to cross the observation area, and C^v and its matching S^v reported as a data point for each simulation. Figure B.14 shows that both the simulation results and the observed data follow a similar trend with a similar dispersion, indicating good agreement.

The above presented tuning of pedestrian-pedestrian, car-car, and car-pedestrian interactions do not provide a valid model for modeling respective

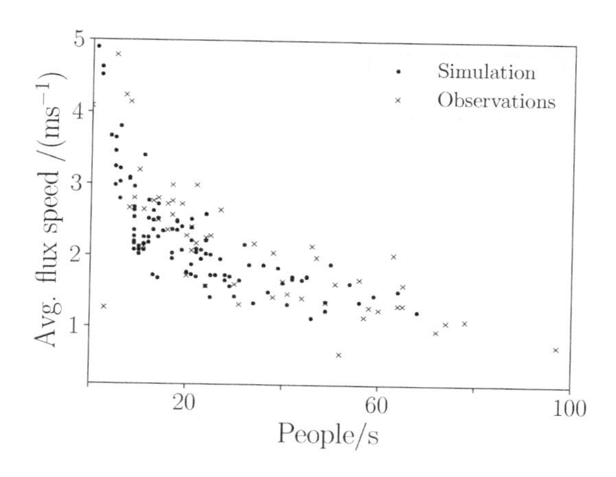

FIGURE B.14: Car-pedestrian interaction validation results.

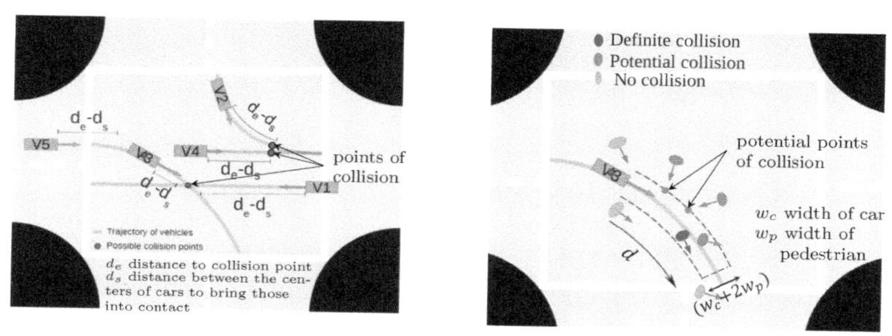

a) Potential collision points of cars on different trajectories, and safe distance to decelerate.

b) Potential collision points of a car with pedestrians.

FIGURE B.15: Illustration potential collision points of cars with other cars and pedestrians.

interactions during a tsunami evacuation since all the observations used were gathered under the ordinary day conditions. Especially, the car-pedestrian interactions were performed during a festive day under no stress condition. Given a similar video recording of people evacuating from a tsunami, etc., same procedures can be used to quantify and tune the model to make agents imitate real evacuees' behaviors.

It is reasonable to assume widespread presence of conditions such as evacuees walking on roads, crossing junctions without respecting traffic signals, broken traffic lights due to the earthquake, etc., during mass evacuations triggered by major natural disasters like an earthquake followed by a tsunami. While above discussed ORCA scheme based collision avoidance enables us to simulate the vehicle-pedestrian interactions on straight roads, some extra functionalities are required to model vehicle-vehicle and vehicle-pedestrian interactions at non-signalized junctions. Modeling the car and pedestrian movements at a signalized junction is relatively simple since the traffic lights provide a centralized control. The absence of centralized controlling makes it significantly difficult to model junctions with non-functioning traffic signals. This subsection briefly explain the modeling of the car-car and car-pedestrian interactions at non-signalized junctions, wherein we assume that

- cars do not deviates from the B-spline trajectories explained in Section 3.3.2
- each car observes the neighboring car's and pedestrian agents' positions, their turn signals, estimates their relative speeds and moving directions
- if a car identifies a potential collision with a car or pedestrian, it avoids the collision by applying comfortable deceleration to maintain a safe distance
- cars give priority to pedestrians

Car-car interactions at non-signalized junctions

The implemented algorithm to resolve collision at non-signalized junctions is based on the simple rule that all the cars obey a common priority. In our current implementation, the car to first arrive at the point of collision is given priority to move uninterrupted by the others.

We classify car-car interaction at a junction of three groups; intersecting trajectories (e.g., V1 and V3 in Fig. B.15a), diverging trajectories, and merging trajectories (e.g., V2 and V4 in Fig. B.15a). The points of collisions are defined as the locations at which two cars of given dimensions on the same or different trajectories come into contact. For the merging trajectories, once the cars V2 and V4 in Fig. B.15a enter the merged road, they maintain a safe gap between them. The safe gap between the two vehicles is calculated considering time to decelerate, including driver's reaction time, and an additional safe distance which is set as the length of a car d_s.

Cars' interaction with pedestrians at non-signalized junctions

Walking is the recommended mode of evacuation in Japan, and hence, we programmed the car agents to give priority to pedestrians. First, a car agent estimates the distance d it would travel before stopping under a comfortable deceleration, and identifies the pedestrian agents occupying its projected curved path of length d (see Fig. B.15b). Also, the car agent calculates which pedestrian agent can enter its projected path of length d and the point of potential collision, according to their relative speeds. Then, the car decelerates to avoid the potential collisions. Although the same basic logic is involved in the car-pedestrian interaction on straight roads, the curved trajectories at junctions make the calculations more involved.

B.3 An automated calibration/optimization tool

Human behaviors change according to the situation, time of the day, area, country, etc. It is highly unlikely to have a model with a unique set of parameters capable of reproducing all the possible situations. The best solution is to provide a versatile tool for calibrating the ABM to the target population using field observations. For this purpose, an automatic calibration tool was developed. The input to the calibration tool is the feasible parameter space for each parameter (e.g., lower and upper bound of each parameter). The evacuation simulation software is then used as a function mapping the parameter space to the observation space. The calibration tool spawns a series of simulations and measure of the disagreement between the evacuation simulator results and observations using a suitable error norm. This way, the problem of finding optimal parameters to mimic given field observations is converted into an error minimization problem. As an example, given a regression on the observed data or a fundamental diagram, $\boldsymbol{y} = r(\boldsymbol{x})$, and n simulation data points (x_i, y_i) in the observation space, the optimal calibration can be evaluated by minimizing a suitable error-norm. We use the L^1-norm (i.e., $\sum_{i=1}^{n} |r(x_i) - y_i|$) for the above presented calibrations of $g^{coll\_av\_ped\_ped}$, $g^{coll\_av\_car\_car}$, and $g^{coll\_av\_car\_ped}$. Another useful measure is L^{∞}-norm which concentrates on the reduction outliers.

Search for the optimal combinations of parameters is a combinatorial optimization problem, and we use heuristic trajectory based approaches to find acceptable parameter set with much lesser computational effort. We implemented two approaches: the first is a progressive search which introduces small random perturbations to the parameters at a small neighborhood of a given parent point, and move to the parameter set with the lowest error. If no neighboring point with a lower error is found after a predefined maximum number of random perturbations, more aggressive perturbations are applied, and the search is restarted at a new point; the second approach is based on simulated annealing, which inverts the former process by starting with more aggressive perturbations and reducing the level of aggressiveness as the search progresses. A schematic view of the search process is seen in Fig. B.16.

This process is computationally intensive as for every trial parameter set a complete simulation must be performed. Furthermore, to assure that the simulation is not being fitted to an outlier/special scenario Monte Carlo simulations are performed with each set of parameters. Monte Carlo simulation and the presence of many independent search fronts increase the parallel invocation of the problem. The automated tool utilizes the Message Passing Interface (MPI) to communicate and spawn the processes.

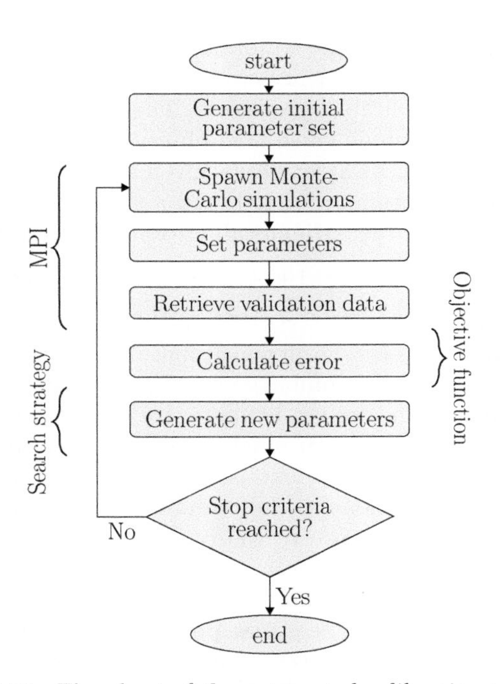

FIGURE B.16: Flowchart of the automated calibration tool.

C

Meta-Modeling Theory

CONTENTS

DOI: 10.1201/9781003149798-C

This appendix summarizes a *meta-modeling theory*. This theory is used to pose various mathematical problems, each providing a suitable approximate solution of a continuum mechanics problem (based on the classical Newtonian mechanics). Here, modeling refers to posing a mathematical problem, and the continuum mechanics problem is an initial-boundary value problem using the wave equation for solids as the governing equation. The essence of the meta-modeling theory is its consistency, i.e., the modeling always provides an approximate solution to the original continuum mechanics problem.

C.1 Structural mechanics from the continuum mechanics viewpoint

Conventional structural mechanics shares various elements with continuum mechanics, such as the displacement-strain relation at the state of infinitesimally small deformation. However, initial-boundary value problems by structural mechanics is not the same as the continuum mechanics problem, thus making the two different. The meta-modeling theory clarifies this point, mathematically proving that the initial-boundary value problems of structural mechanics can be derived from the continuum mechanics problem. The derivation solely uses the mathematical approximations[1] and does not make any physical assumptions.

From the physical viewpoints, continuum mechanics relies on three elements, kinematics, dynamics and material properties, which are expressed in the equations with functions of displacement, strain, and stress. The wave equation is mathematically derived from these equations, and an initial-boundary value problem is posed for a continuum body, denoted by V, when suitable initial and boundary conditions are provided for V. The initial-boundary-value problem is further converted into a variational problem. For simplicity, assuming linear elasticity and infinitesimally small deformation, we can pose a variational problem which uses the following functional for a displacement function, $\boldsymbol{u}$:

$$\mathcal{L}[\boldsymbol{u}] = \mathcal{K}[\boldsymbol{v}] - \mathcal{P}[\boldsymbol{\epsilon}], \tag{C.1}$$

where $\boldsymbol{v}$ and $\boldsymbol{\epsilon}$ are the velocity and strain, respectively, defined as $\boldsymbol{v} = \dot{\boldsymbol{u}}$ and $\boldsymbol{\epsilon} = \mathrm{sym}\{\nabla \boldsymbol{u}\}$ with $\dot{(.)}$ and $\nabla(.)$ being the temporal derivative and gradient

[1] While mathematical approximations do not have to be validated, physical assumptions ought to be validated based on experimental data, field observation or actual measurement. Since bad mathematical approximations result in inaccurate approximate solutions, only good mathematical approximations are used, which also exhibit limited ability to produce accurate approximate solutions.

for a function (.), respectively, and sym representing the symmetric part, and

$$\mathcal{K}[\boldsymbol{v}] = \int_V \frac{1}{2}\rho \boldsymbol{v} \cdot \boldsymbol{v} \, \mathrm{d}v, \quad \mathcal{P}[\boldsymbol{\epsilon}] = \int \frac{1}{2}\boldsymbol{\epsilon} : \boldsymbol{c} : \boldsymbol{\epsilon} \, \mathrm{d}v. \tag{C.2}$$

with ρ and $\boldsymbol{c}$ being the density and elasticity tensor; $\cdot$ and $:$ are the inner product and second-order contraction, respectively. By definition, $\mathcal{K}$ and $\mathcal{L}$ are the kinematic and potential energy of V, respectively. Since $\mathcal{L}$ is[2] a Lagrangian, we seek $\boldsymbol{u}$ such that $\mathcal{L}$ is stationalized.

The meta-modeling theory uses a stress function, $\boldsymbol{\sigma}$, in addition to a displacement function, $\boldsymbol{u}$, and employs the following functional for $\boldsymbol{u}$ and $\boldsymbol{\sigma}$:

$$\mathcal{L}^*[\boldsymbol{u}, \boldsymbol{\sigma}] = \mathcal{K}[\boldsymbol{v}] - \mathcal{P}^*[\boldsymbol{\epsilon}, \boldsymbol{\sigma}], \tag{C.3}$$

where

$$\mathcal{P}^*[\boldsymbol{\epsilon}, \boldsymbol{\sigma}] = \int \boldsymbol{\sigma} : \boldsymbol{\epsilon} - \frac{1}{2}\boldsymbol{\sigma} : \boldsymbol{c}^{-1} : \boldsymbol{\sigma} \, \mathrm{d}v, \tag{C.4}$$

with $\boldsymbol{c}^{-1}$ being the inverse of $\boldsymbol{c}$. This $\mathcal{L}^*$ is equivalent with $\mathcal{L}$, since $\delta \int \mathcal{L}^* \, \mathrm{d}t = 0$ yields

$$\int -\delta\boldsymbol{\sigma} : \left(\boldsymbol{\epsilon} - \boldsymbol{c}^{-1} : \boldsymbol{\sigma}\right) - \delta\boldsymbol{u} \cdot \left(\rho\ddot{\boldsymbol{u}} - \boldsymbol{\nabla} \cdot \boldsymbol{\sigma}\right) \mathrm{d}v = 0,$$

and $\boldsymbol{u}$ and $\boldsymbol{\sigma}$ must satisfy $\boldsymbol{\sigma} = \boldsymbol{c} : \boldsymbol{\epsilon}$ and $\rho\,\ddot{\boldsymbol{u}} - \boldsymbol{\nabla} \cdot \boldsymbol{\sigma} = \boldsymbol{0}$. Substituting the first equation into the second equation, we derive

$$\rho\,\ddot{\boldsymbol{u}} - \boldsymbol{\nabla} \cdot (\boldsymbol{c} : \boldsymbol{\nabla}\boldsymbol{u}) = \boldsymbol{0}, \tag{C.5}$$

which coincides with the wave equation; see Eq. (2.14). Therefore, $\mathcal{L}^*[\boldsymbol{u}, \boldsymbol{\sigma}]$ is equivalent with $\mathcal{L}[\boldsymbol{u}]$ in the sense that $\delta\mathcal{L}^* \, \mathrm{d}t = 0$ and $\delta \int \mathcal{L} \, \mathrm{d}t = 0$ give the identical solution.

C.2 Derivation of governing equation of bar problem

Simplest modeling of structural mechanics poses a bar problem, giving an approximate solution of a continuum body like a bar, as per the meta-modeling theory. It should be pointed out that one-dimensional stress-strain relation is usually assumed as the bar material property, i.e., $\sigma = E\,\epsilon$, where σ and ϵ are normal components of stress and strain tensors in the longitudinal direction, and E is Young's modulus. However, this physical assumption of the material property is not validated in the experiment, and the Poisson effect, involving

[2] We can define the Hamiltonian of V as $\mathcal{K} + \mathcal{L}$. It is the Hamiltonian that corresponds to the conservation of energy. The Lagrangian of V, such as $\mathcal{L}$, is of the dimension of energy but does not correspond to the conservation of energy.

bar shrinking in the transverse direction when pulled in the longitudinal direction. Notably, the material property does not depend on the configuration of the body.

The meta-modeling theory uses mathematically approximated functions for the variational problem of $\mathcal{L}^*$. For a bar, employing a Cartesian coordinate system, (x_1, x_2, x_3), with the x_1-axis being the bar longitudinal direction, we observe that

$$|u_1| \gg |u_2|,\ |u_3|, \quad |\sigma_{11}| \gg |\sigma_{22}|,\ |\sigma_{33}|,\ |\sigma_{23}|,\ |\sigma_{31}|,\ |\sigma_{12}|.$$

Also, for u_1 and σ_{11}, the dependence on x_2 or x_3 is negligibly smaller compared with that on x_1. Hence, we use $\boldsymbol{u}$ and $\boldsymbol{\sigma}$ of the following form:

$$u_1 = u(x_1, t), \quad \sigma_{11} = \sigma(x_1, t), \tag{C.6}$$

with other components of $\boldsymbol{u}$ and $\boldsymbol{\sigma}$ being zero. This setting of $\boldsymbol{u}$ and $\boldsymbol{\sigma}$ is highly important for the meta-modeling theory since it specifies the mathematical approximation of the displacement and stress functions for the bar.

Substituting Eq. (C.6) into Eq. (C.3), we compute $\mathcal{L}^*$ as

$$\mathcal{L}^* = \int \frac{1}{2}\rho\, A\, \dot{u}^2 - A\left(u'\sigma - \frac{1}{2}\frac{\sigma^2}{E}\right) \mathrm{d}x_1, \tag{C.7}$$

where $A = \int \mathrm{d}x_2 \mathrm{d}x_3$ is the cross-sectional area and prime is the derivative with respect to x_1. While $\boldsymbol{c}$ in Eq. (C.3) includes Poisson's ratio, ν, wherein ν is dropped in Eq. (C.7). Since $\partial \int \mathcal{L}^*\, \mathrm{d}t = 0$ with respect to σ yields $\sigma = Eu'$, substituting this σ into Eq. (C.7) and taking variation of the resulting $\mathcal{L}^*$ with respect to u, we can readily derive an initial-boundary value problem for u, the governing equation of which is

$$\rho\, A\, \ddot{u} - (EAu')' = 0. \tag{C.8}$$

This equation coincides with the governing equation of a bar problem of structural mechanics. According to the meta-modeling theory, the bar problem can be a mathematical approximation of the continuum mechanics problem that uses the wave equation of Eq. (C.5) as the governing equation.

C.3 Derivation of governing equation of beam problem

Similar to the bar problem, we can also derive the governing equation of a beam problem, employing the Cartesian coordinate system considered in the preceding section for a beam as well. Regarding the beam displacement and stress, we observe that

$$|u_3| \gg |u_1|,\ |u_2|, \quad |\sigma_{11}| \gg |\sigma_{22}|,\ |\sigma_{33}|,\ |\sigma_{23}|,\ |\sigma_{31}|,\ |\sigma_{12}|.$$

Here, for simplicity, the x_3-axis is taken as the transverse direction of the bar wherein the beam bends. It should be recalled that σ_{ij}'s except σ_{11} are smaller due to the traction free boundary conditions on the side surfaces of the beam. We thus use $\boldsymbol{u}$ and $\boldsymbol{\sigma}$ of the following form:

$$u_3 = w(x_1, t), \quad u_1 = -x_3 w'(x_1, t), \quad \sigma_{11} = \sigma(x_1, x_3, t), \tag{C.9}$$

with other components of $\boldsymbol{u}$ and $\boldsymbol{\sigma}$ being zero. It is necessary to use u_1 of the above form to vanish the shear strain component, $\epsilon_{13} = \frac{1}{2}(\frac{\partial u_1}{\partial x_3} + \frac{\partial u_3}{\partial x_1})$. The longitudinal displacement component of the beam, u_1, is more complicated than that of the bar, while the transverse displacement component of the beam, u_3, is the same as the longitudinal component of the bar.

Substituting Eq. (C.9) into Eq. (C.3), we compute $\mathcal{L}^*$ as

$$\mathcal{L}^* = \int \frac{1}{2}\rho\,\dot{w}^2 - \Big(- x_3 w''\sigma - \frac{1}{2}\frac{\sigma^2}{E}\Big)\,\mathrm{d}v. \tag{C.10}$$

Like Eq. (C.7), ν is dropped in Eq. (C.10) because of $\boldsymbol{\sigma} : \boldsymbol{c}^{-1} : \boldsymbol{\sigma} = \frac{\sigma^2}{E}$ for $\boldsymbol{\sigma}$ of the approximated form. Computing variation with respect to σ, we have

$$\int \delta\sigma\Big(x_3 w'' - \frac{\sigma}{E}\Big)\,\mathrm{d}v\,\mathrm{d}t = 0,$$

from which we obtain $\sigma = -E x_3 w''$. The longitudinal normal stress component of the bar becomes complicated, but it can be expressed in terms of w. Thus, substituting this σ into Eq. (C.10) and taking variation of the resulting $\mathcal{L}^*$ with respect to w, we have

$$-\int \delta w\Big(\rho\ddot{w} + (E\,x_3^2 w'')''\Big)\,\mathrm{d}v\,\mathrm{d}t = 0.$$

We can derive an initial-boundary value problem for w, the governing equation of which is

$$\rho\,A\,\ddot{w} - (EIw'')'' = 0, \tag{C.11}$$

where $I = \int x_3^2\,\mathrm{d}x_2 \mathrm{d}x_3$ is the second moment of inertia. Equation (C.11) coincides with the governing equation of the beam problem. Like the bar problem, according to the meta-modeling theory, the beam problem of structural mechanics can be regarded a mathematical approximation of the continuum mechanics problem.

It should be emphasized that the governing equation of a different form, Eq. (C.8) or (C.11), can provide an approximate solution of Eq. (C.5) or the wave equation. The approximate solution is close to the exact solution, if the target body is like a bar (or a beam); the traction free boundary conditions on the side surfaces are important since we can approximate that only the normal component in the longitudinal direction is a non-zero stress component. Thus, the approximate solution of the beam problem is applicable to a body like a bar which satisfies the traction-free boundary conditions on the side surfaces.

C.4 Derivation of governing equation of torsional bar problem

The last example of the meta-modeling theory is a torsional bar problem, which is rarely studied in earthquake engineering. We employ a bar as the target body and torsion as the external force, and consider the same Cartesian coordinate system as the bar problem. Regarding the displacement, we observe that components in the transverse directions, u_2 and u_3, are produced from torsional loading, which induces the component in the longitudinal direction, u_1, as well. Regarding the stress of the torsional bar, we observe that

$$|\sigma_{11}|,\ |\sigma_{12}|,\ |\sigma_{13}| \gg |\sigma_{22}|,\ |\sigma_{33}|,\ |\sigma_{23}|.$$

On the side surface, the shear stress components in the longitudinal direction, σ_{12} and σ_{13}, must vanish. However, at the cross-section, σ_{12} and σ_{13} are produced due to torsional loading. We thus use $\boldsymbol{u}$ and $\boldsymbol{\sigma}$ of the following form:

$$\begin{aligned} &u_1 = u(x_1,x_2,x_3,t), \quad u_2 = -\phi(x_1,t)x_3, \quad u_3 = \phi(x_1,t)x_2, \\ &\sigma_{11} = \sigma(x_1,x_2,x_3,t), \quad \sigma_{12} = \tau_2(x_1,x_2,x_3,t), \\ &\sigma_{13} = \tau_3(x_1,x_2,x_3,t), \end{aligned} \tag{C.12}$$

with other stress components being zero. As is observed, ϕ is the rotation of the torsional deformation, an essential approximation for the torsional bar.

Substituting Eq. (C.12) into Eq. (C.3), we compute $\mathcal{L}^*$ as

$$\begin{aligned} \mathcal{L}^* = \int \frac{1}{2}\rho(\dot{u}^2 + (x_2^3 + x_3^2)\dot{\phi}^2) - \Big(\tau_2(u_{,2} - x_3\phi_{,1}) + \tau_3(u_{,3} + x_2\phi_{,1}) \\ -\frac{1}{2}\frac{\sigma^2}{E} - \frac{1}{4}\frac{\tau_2^2+\tau_3^2}{G}\Big)\, \mathrm{d}v, \end{aligned} \tag{C.13}$$

where G is the shear modulus, and an index following comma represents the partial differentiation, i.e., $(.)_{,i} = \frac{\partial (.)}{\partial x_i}$. Computing variation of this $\mathcal{L}^*$ with respect to σ, τ_2, and τ_3, we have

$$\int \delta\sigma\Big(\frac{\sigma}{E} - u_{,1}\Big) + \frac{\delta\tau_2}{2}\Big(\frac{\tau_2}{G} - u_{,2} + x_3\phi_{,1}\Big) + \frac{\delta\tau_3}{2}\Big(\frac{\tau_3}{G} - u_{,3} - x_2\phi_{,1}\Big)\, \mathrm{d}v\, \mathrm{d}t = 0,$$

from which we obtain $\sigma = E\,u_{,1}$, $\tau_2 = G(u_{,2} - x_3\phi'_{,1})$, and $\tau_3 = G(u_{,3} + x_2\phi_{,1})$. Substituting these σ, τ_2, and τ_3 into Eq. (C.13) and taking variation of the resulting $\mathcal{L}^*$ with respect to u and ψ, we have

$$\begin{aligned} -\int \delta u\Big(\rho\ddot{u} - (E\,u_{,1})_{,1}\Big) + \delta\phi\Big(\rho(x_2^2 + x_3^2)\ddot{\phi} - G(x_2^2 + x_3^2)\phi_{,11}) \\ -x_3 u_{,12} + x_2 u_{,13}\Big)\, \mathrm{d}v\, \mathrm{d}t = 0. \end{aligned}$$

Therefore, we can derive a coupled initial-boundary value problem for u and ϕ, the governing equations of which are

$$\begin{cases} M\,\ddot{u} - (EIu_{,1})_{,1} - G(u_{,22} + u_{,33}) = 0, \\ \rho(x_2^2 + x_3^2)\ddot{\phi} - G\Big((x_2^2 + x_3^2)\phi_{,11} - x_3u_{,2} + x_2u_{,3}\Big) = 0. \end{cases} \tag{C.14}$$

This set coincide with the governing equations of the torsional bar. If we further approximate $u_{,1} = 0$, based on the observation of $|u_{,1}| \ll |u_{,2}|, |u_{,3}|$ and consider the quasi-static state, then, Eq. (C.14) is reduced[3] to two-dimensional problems of $u_{,22} + u_{,33} = 0$ and $I\phi_{,11} = 0$ at the cross-section. As is observed, u and ϕ are decoupled in these equations.

[3]An additional approximation of $\int x_3u_{,2} - x_2u_{,3}\mathrm{d}x_2\mathrm{d}x_3 = 0$ is made in deriving the second equation of Eq. (C.14).

D

Mathematical Treatment of Soil-Structure Interaction

CONTENTS

DOI: 10.1201/9781003149798-D

D.1 Soil-structure interaction effects

In earthquake engineering, the evaluation of soil-structure-interaction has been a critical, long-term issue in considering the seismic safety of a structure. Soil-structure interaction can significantly affect the structural seismic responses, since the response changes depending on the soil on which the structure is based. The effects of soil-structure interaction depend on input ground motion as well. Furthermore, the natural frequency of a structure is influenced by the soil conditions, as will be shown later.

Continuum mechanics can accurately estimate the mechanical behavior of a solid or fluid body if it is much larger than the molecules or atoms that constitute the body. It leads to wave equation, a set of coupled four-dimensional partial differential equations for three functions which serve as the governing equation. The wave equation is simple since it does not include any terms of interactions; it is a differential equation for the temporal and spatial derivatives of a displacement function at a single time and spatial point. The mechanical interaction between two distinct bodies can be accurately evaluated by solving the wave equation for the two bodies. It should be recalled that the Newton's second law is expressed in the form of a temporal differential equation at one instance. This equation does not include any terms of the past or future. However, the effects of a past event on the future can be evaluated solely by solving the equation. The wave equation in continuum mechanics plays the same role as the Newton's second law.

According to continuum mechanics, soil-structure interaction can be precisely evaluated by solving the wave equation for the soil and the structure (with given configuration and material properties), provided that the initial and boundary conditions as well as the interface conditions between them are prescribed. In other words, there could be multiple sources of errors in the initial-boundary value problem of the wave equation. For instance, inputting the same ground motion on the bottom of an analysis model may not be the most accurate treatment. Moreover, proper treatment of boundary conditions at an artificial boundary is not fully achieved when a finite domain is used for the numerical analysis. The interface conditions could be more complicated than the simple interface condition of perfect bonding, when a slip or detachment occurs at the interface.

D.2 Formulation of soil spring

Instead of posing correct initial, boundary, and interface conditions, simpler treatments have been developed to estimate the soil-structure interaction. When it starts moving subjected to the ground motion, a structure exerts reactive forces on the soil, and the soil deforms accordingly. The displacement response of the soil to the structure's force can be simply modeled using a spring. This is an idea of introducing soil spring. The soil spring is a simple tool to estimate the effects of soil-structure interaction on the structural seismic response. It should be pointed out that the spring constant of the soil spring can be estimated for an ideal setting such as a flat plane or a homogeneous body. When a concentrated force is applied at a point on the surface of a half plane, a closed-form solution or a solution in the form of series expansion can be obtained solving the wave equation. A solution in a series expansion form can be obtained for a more complicated setting of the half space.

For a given set of structure and soil, it is possible to construct a simple model which consists of mass and spring, using a Lagrangian of continuum mechanics. The key feature of this approach is that the constructed mass-spring model is consistent with the original continuum mechanics problem of the structure and soil. That is, similar to the meta-modeling theory explained in Appendix C, this approach provides an approximate solution of the continuum mechanics problem.

As the simplest example, we consider a linearly elastic body, which is studied in Subsection 2.3.2. A Lagrangian is formulated for a system of a structure and soil, denoted by S and G, respectively; it is a functional for a displacement function $\boldsymbol{u}$ and given as

$$\mathcal{L}[\boldsymbol{u}] = \int_{S+G} \frac{1}{2}\rho \boldsymbol{v} \cdot \boldsymbol{v} - \frac{1}{2}\boldsymbol{\epsilon} : \boldsymbol{c} : \boldsymbol{\epsilon}\, \mathrm{d}v, \tag{D.1}$$

where $\boldsymbol{v} = \dot{\boldsymbol{u}}$ and $\boldsymbol{\epsilon} = \mathrm{sym}\boldsymbol{\nabla}\boldsymbol{u}$ are the velocity and strain, with $\dot{(.)}$ and $\boldsymbol{\nabla}(.)$ being the temporal derivative and gradient and sym representing the symmetric part, ρ and $\boldsymbol{c}$ are density and elasticity tensor, and $\cdot$ and $:$ are the inner product and second-order contraction, respectively. It should be recalled that there are no terms of interaction in $\mathcal{L}$ of Eq. (D.1).

To estimate the soil-structure interaction, we consider suitable mathematical approximations for $\boldsymbol{u}$, and seek to construct a consistent mass-spring model with soil spring. Attention must be paid to the condition on the interface between S and G, denoted by I. In a conventional approach, I is modeled as an infinitesimally thin and rigid body plate, to enforce the point-wise continuity of displacement and traction across I. Hereafter, we consider the simple case that input ground motion is horizontal, given as $g\boldsymbol{\psi}$ where g is the amplitude of input ground motion and $\boldsymbol{\psi}$ is a horizontal unit vector. This $g\boldsymbol{\psi}$ is inserted in I. It should be noted that the symbols used here are slightly different from those in Subsection 2.3.2.

As the first construction step of a consistent mass-spring model with a soil spring, we use the following $\boldsymbol{u}$ to computation $\mathcal{L}$:

$$\boldsymbol{u}(\boldsymbol{x}, t) = g(t)\boldsymbol{\psi} + u(t)\boldsymbol{\phi}(\boldsymbol{x}) \quad \text{in } S, \tag{D.2}$$

where $g\boldsymbol{\psi}$ is a rigid body translation and $u\boldsymbol{\phi}$ represents the first mode of the structural response, with u and $\boldsymbol{\phi}$ being the amplitude and shape, respectively. By definition, $\boldsymbol{\phi}$ satisfies

$$\rho(\boldsymbol{x})\,\omega^2\boldsymbol{\phi}(\boldsymbol{x}) + \boldsymbol{\nabla}\cdot(\boldsymbol{c}(\boldsymbol{x}) : \boldsymbol{\nabla}\boldsymbol{\phi}(\boldsymbol{x})) = \boldsymbol{0},$$

where ω is the natural frequency of the first mode; ρ and $\boldsymbol{c}$ are considered as a spatial function since various materials are used for S. Substituting Eq. (D.2) into Eq. (D.1), we compute $\mathcal{L}$ as

$$\mathcal{L} = \frac{1}{2}m^{\psi\psi}\dot{g}^2 + m^{\psi\phi}\dot{g}\dot{u} + \frac{1}{2}m^{\phi\phi}\dot{u}^2 - \frac{1}{2}k^{\phi\phi}u^2, \tag{D.3}$$

where $m^{\phi\phi}$ is defined as $m^{\phi\phi} = \int_S \rho\boldsymbol{\phi}\cdot\boldsymbol{\phi}\,\mathrm{d}v$ with $m^{\psi\psi}$ and $m^{\psi\phi}$ being defined in the same manner as $m^{\phi\phi}$, and $k^{\phi\phi}$ is defined as $k^{\phi\phi} = \int_S \boldsymbol{\nabla}\boldsymbol{\phi}\boldsymbol{c} : \boldsymbol{\nabla}\boldsymbol{\phi}\,\mathrm{d}v$. It should be noted that $m^{\phi\phi}$ and $k^{\phi\phi}$ satisfy $\omega^2 m^{\phi\phi} - k^{\phi\phi} = 0$, since $\boldsymbol{\phi}$ is the mode shape.

Since $\boldsymbol{\psi}$ is the unit vector, we compute $m^{\psi\psi} = \int_S \rho\,\mathrm{d}v$, which is the total mass of S. Standardizing $\boldsymbol{\phi}$ to satisfy $m^{\phi\phi} = m^{\psi\psi}$, we can derive the following equation for u from $\delta\int\mathcal{L}\,\mathrm{d}t = 0$:

$$m^{\phi\phi}\ddot{u} + k^{\phi\phi}u = -m^{\psi\phi}\ddot{g}, \tag{D.4}$$

which coincide with the governing equation of the mass-spring model with the mass $m^{\phi\phi}$ and the spring constant $k^{\phi\phi}$; strictly speaking, $m^{\psi\phi}$ does not satisfy $m^{\psi\phi} = m^{\psi\psi}$. The amplitude of the first mode, u, gives an approximate displacement at a certain point in S, $\boldsymbol{x}^{\mathrm{obs}}$ which satisfies $|\boldsymbol{\phi}(\boldsymbol{x}^{\mathrm{obs}})| = 1$, or an average displacement for a certain part of S.

At the next construction step of a consistent mass-spring model with a soil spring, we consider the displacement in G in addition to the displacement in S. That is,

$$\boldsymbol{u}(\boldsymbol{x},t) = \begin{cases} (g(t)+U(t))\boldsymbol{\psi} + u(t)\boldsymbol{\phi}(\boldsymbol{x}) & \text{in } S, \\ g(t)\boldsymbol{\Psi}^0(\boldsymbol{x}) + U(t)\boldsymbol{\Psi}^1(\boldsymbol{x}) & \text{in } G. \end{cases} \tag{D.5}$$

Here, U is an additional translation of I; $\boldsymbol{\Psi}^0$ is the displacement field in G caused by the input ground motion[1] in the absence of S; and $\boldsymbol{\Psi}^1$ is the displacement field in G induced by U. Comparing Eq. (D.5) with Eq. (D.2), we can see the following two differences: 1) U is added to the displacement function in S; and 2) U generates $\boldsymbol{\Psi}^1$ in G. By definition, U is added to the translation of I induced by the input ground motion g, and U produces additional rigid-body motion to S and ground deformation $\boldsymbol{\Phi}^1$ in S. Substituting Eq. (D.5) into Eq. (D.1), we compute $\mathcal{L}$ as

$$\begin{aligned}\mathcal{L} = {} & \frac{1}{2}m^{\psi\psi}(\dot{g}+\dot{U})^2 + m^{\psi\phi}(\dot{g}+\dot{U})\dot{u} + \frac{1}{2}m^{\phi\phi}\dot{u}^2 - \frac{1}{2}k^{\phi\phi}u^2 + \frac{1}{2}M^{00}\dot{g}^2 \\ & + M^{01}\dot{g}\dot{U} + \frac{1}{2}M^{11}\dot{U}^2 - \frac{1}{2}K^{00}g^2 - k^{01}gU - \frac{1}{2}K^{11}U,\end{aligned} \tag{D.6}$$

[1] In general, $\boldsymbol{\Psi}^0$ and $\boldsymbol{\Psi}^1$ ought to depend on t. Here, for simplicity, we neglect the dependence of $\boldsymbol{\Psi}^0$ and $\boldsymbol{\Psi}^1$ on t. A more accurate approximate solution is obtained if we use time-dependent $\boldsymbol{\Psi}^0$ even though it is not easy to find a suitable $\boldsymbol{\Psi}^0$.

where $M^{\alpha\beta} = \int_G \rho \boldsymbol{\Psi}^\alpha \cdot \boldsymbol{\Psi}^\beta \,\mathrm{d}v$ and $K^{\alpha\beta} = \int_G \nabla\boldsymbol{\Psi}^\alpha : \boldsymbol{c} : \nabla\boldsymbol{\Psi}^\beta \,\mathrm{d}v$.

Like the previous step, we compute the vanishing of the variation, $\delta \int \mathcal{L}\,\mathrm{d}t = 0$, and derive the following set of differential equations:

$$\begin{cases} m^{\phi\phi}\ddot{u} + k^{\phi\phi}u = -m^{\psi\phi}(\ddot{g} + \ddot{U}), \\ M^{11}\ddot{U} + K^{11}U = -m^{\phi\phi}(\ddot{g} + \ddot{U}) - m^{\psi\phi}\ddot{u} - M^{01}\ddot{g} - K^{01}g. \end{cases}$$

All the parameters in the equations are computed when $\boldsymbol{\Psi}^0$ and $\boldsymbol{\Psi}^1$ are given. Now, if we assume that the second equation is solved for $\ddot{U}$ and that $m^{\psi\phi}\ddot{U}$ is expressed in terms of u as K^*u, then, we can rewrite the first equation as

$$m^{\phi\phi}\ddot{u} + k^{\phi\phi}u = -m^{\psi\phi}\ddot{g} - K^*u, \tag{D.7}$$

which coincides with the governing equation of a mass-spring model with a sway soil spring.

The value of the spring constant, K^*, changes depending on the choice of $\boldsymbol{\Psi}^1$. Usually, the ground deformation induced by a harmonic loading of I is used for $\boldsymbol{\Psi}^1$, when the period of the loading is suitably presumed. It is certainly true that the sway soil spring can provide a good approximate solution for a certain class of input ground motions if a suitable period is chosen for $\boldsymbol{\Psi}^1$. However, it does not mean that the sway soil spring that is determined using this $\boldsymbol{\Psi}^1$ can always provide a good approximate solution.

It is possible to construct a multiple mass-spring model with sway soil springs, which consists of a few masses and springs. For instance, we consider the case wherein a set of dynamic modes of S and a set of harmonic ground deformations induced by I are given. We use the following displacement function:

$$\boldsymbol{u}(\boldsymbol{x},t) = \begin{cases} (g(t) + \sum_\alpha U^\alpha(t))\boldsymbol{\psi} + \sum_\alpha u^\alpha(t)\boldsymbol{\phi}^\alpha(\boldsymbol{x}) & \text{in } S, \\ g(t)\boldsymbol{\Psi}^0(\boldsymbol{x}) + \sum_\alpha U^\alpha(t)\boldsymbol{\Psi}^\alpha(\boldsymbol{x}) & \text{in } G, \end{cases} \tag{D.8}$$

where u^α and $\boldsymbol{\phi}^\alpha$ are the amplitude and shape of the α-th mode of S, and U^α and $\boldsymbol{\psi}^\alpha$ are the amplitude and ground deformation induced by the α-th harmonic loading of I. Substituting Eq. (D.8) into Eq. (D.1), we compute $\mathcal{L}$ as

$$\mathcal{L} = \frac{1}{2}m^{\psi\psi}\left(\sum_\alpha \dot{U}^\alpha\right)^2 + \sum_\alpha m^{\psi\alpha}\left(\sum_\beta \dot{U}^\beta\right)\dot{u}^\alpha + \frac{1}{2}m^{\alpha\alpha}(\dot{u}^\alpha)^2 - \frac{1}{2}k^{\alpha\alpha}(u^\alpha)^2$$
$$+ \sum_{\alpha,\beta}\frac{1}{2}M^{\alpha\beta}\dot{U}^\alpha\dot{U}^\beta - \frac{1}{2}K^{\alpha\beta}U^\alpha U^\beta. \tag{D.9}$$

Here, for simplicity, g is denoted by U^0, and $g + \sum U^\alpha$ is replaced by the expression of $\sum U^\alpha$. We can arrive at the following set of differential equations from $\delta \int \mathcal{L}\,\mathrm{d}t = 0$:

$$\begin{cases} m^{\alpha\alpha}\ddot{u}^\alpha + k^{\alpha\alpha}u^\alpha = -m^{\psi\alpha}\sum_\beta \ddot{U}^\beta, \\ \sum M^{\alpha\beta}\ddot{U}^\beta + \sum K^{\alpha\beta}U^\beta = -m^{\psi\psi}\sum_\beta \ddot{U}^\beta - \sum_\beta m^{\psi\beta}\ddot{u}^\beta. \end{cases}$$

Now, we assume that the second equation is solved for $\{U^\alpha\}$ and that $m^{\psi\alpha}\ddot{U}^\beta$ in the first equation is expressed in terms of $\{u^\alpha\}$ as $m^{*\alpha}\ddot{g}+K^{*\alpha}u^\alpha$, then, we can rewrite the first equation as

$$m^{\alpha\alpha}\ddot{u}^\alpha + k^{\alpha\alpha}u^\alpha = m^{*\alpha}g + K^{*\alpha}u^\alpha. \tag{D.10}$$

This equation coincides with the governing equation of the α-th mode, influenced by the sway spring with the spring constant $K^{*\alpha}$.

In addition to a sway spring, we can include a rocking spring for the soil spring. For simplicity, we go back to the first step and solely consider the first mode of S and a function of translation and rotation of I. That is, we use the following displacement function:

$$\boldsymbol{u}(\boldsymbol{x},t) = \begin{cases} (g(t)+U(t))\boldsymbol{\psi} + \Theta(t)\boldsymbol{\xi}(\boldsymbol{x}) + u(t)\boldsymbol{\phi}(\boldsymbol{x}) & \text{in } S, \\ g(t)\boldsymbol{\Psi}^0(\boldsymbol{x},t) + U(t)\boldsymbol{\Psi}^1(\boldsymbol{x}) + \Theta(t)\boldsymbol{\xi}(\boldsymbol{x}) & \text{in } G, \end{cases} \tag{D.11}$$

where Θ is the amplitude of rotation of I, $\boldsymbol{\xi}$ is a linear function of $\boldsymbol{x}$, which corresponds to the rigid-body rotation of S induced by the rotation of I, and $\boldsymbol{\Xi}$ is the ground deformation induced by the rotation of I. Substituting Eq. (D.11) into Eq. (D.1), we compute $\mathcal{L}$ as

$$\begin{aligned}\mathcal{L} &= \frac{1}{2}m^{\psi\psi}(\dot{g}+\dot{U})^2 + \frac{1}{2}i^{\xi\xi}\dot{\Theta}^2 + m^{\psi\phi}(\dot{g}+\dot{U})\dot{u} + j^{\xi\phi}\dot{\Theta}\dot{u} - k^{\phi\phi}u^2 \\ &+\frac{1}{2}M^{00}(\dot{g})^2 + \frac{1}{2}M^{11}\dot{U}^2 + \frac{1}{2}I^{11}\dot{\Theta}^2 + M^{01}\dot{g}\dot{U} - \frac{1}{2}K^{00}g^2 - \frac{1}{2}K^{11}U^2 \\ &-\frac{1}{2}R^{11}\Theta^2 - K^{01}gU,\end{aligned} \tag{D.12}$$

where $i^{\xi\xi}$ and $j^{\xi\phi}$ are the integrations over S similar to $m^{\phi\phi}$, and I^{11} and R^{11} are the integrations over G similar to M^{11} and K^{11}. We can derive the following equation from $\delta\int\mathcal{L}\,\mathrm{d}t=0$:

$$\begin{cases} m^{\phi\phi}\ddot{u} + k^{\phi\phi}u = -m^{\psi\phi}(\ddot{g}+\ddot{U}) - j^{\xi\phi}\ddot{\Theta}, \\ M^{11}\ddot{U} + K^{11}U = -m^{\psi\psi}(\ddot{g}+\ddot{U}) - m^{\psi\phi}\ddot{u} - M^{01}\ddot{g} - K^{01}g, \\ I^{11}\ddot{\Theta} + R^{11}\Theta = -j^{\xi\phi}\ddot{u}. \end{cases}$$

For instance, if we assume that $\ddot{U}$ and $\ddot{\Theta}$ in the right side of the first equation are expressed in terms of u, then, we can introduce a rocking spring in addition to a sway spring to the mass-spring model of mass $m^{\phi\phi}$ and spring constant $k^{\phi\phi}$. It should be pointed out that suitable approximations must be made to solve the second and third equations and to express $\ddot{U}$ and $\ddot{\Theta}$ in terms of u. We can use Θ as an unknown function in addition to u, since Θ corresponds to the rigid body rotation of S. In this case, a set of coupled equations for u and Θ must be approximately derived from the above equations, expressing U in terms of u and Θ.

D.3 Applicability and limitation of soil spring

It should be emphasized that a mass-spring model with soil spring is used to estimate soil-structure interaction, when it is difficult to construct a high-fidelity analysis model for a structure and soil, as well as to correctly prescribe the initial, boundary, and interface conditions. It is sufficient if the mass-spring model can provide a suitably accurate estimate of soil-structure interaction. To this end, the following two conditions must be met: 1) suitable functions need to be used to compute soil spring constants; and 2) suitable approximations are made to relate the translation (or rotation) of the interface to the displacement of the structure. These two conditions are important, because the applicability of the spring constant is limited to a certain class of input ground motions. It should be also pointed out that using a larger number of soil springs may not necessarily contribute to the increase in the accuracy of the evaluation of the soil-structure interaction. It is much easier to use a high-fidelity analysis model for the structure and soil, which can be applied to a wide range of ground motions, considering suitable boundary and interface conditions for a more accurate evaluation of the soil-structure interaction.

In the formulation of the soil spring presented above, we must pay attention the following two approximations: 1) an approximate displacement function considering the structural response and the interface movement which are used in the functional $\mathcal{L}$; and 2) assumptions which are made to determine soil spring constants, suitably solving the equations derived from $\int \mathcal{L}\,\mathrm{d}t = 0$. The first approximation is used in the meta-modeling theory, but the second approximation is unique to the soil spring formulation. Considerable approximations must be made to determine the soil spring constants, compared with setting an approximate displacement function; no approximations are needed to compute $\mathcal{L}$ and derive the governing equation from $\int \mathcal{L}\,\mathrm{d}t = 0$. Strictly speaking, the governing equations derived from $\mathcal{L}$ are a coupled equations for the structural response and the interface movement, and hence it is difficult to solve the equations for the interface only to determine soil springs which relate the structural response to the interface movement. A suitable approximate displacement function is important to determine the soil spring using $\mathcal{L}$, but making suitable approximations in solving the derived equations is more important to relate the structural response to the interface movement, which corresponds to the soil spring.

In closing this appendix, we summarize the procedure of formulating soil spring using a Lagrangian of continuum mechanics. We start from the introduction of the interface movement into a displacement function used in the Lagrangian. Using a suitable approximate displacement function, we can derive a coupled problem for the structural response and the interface movement from the stationary condition of the Lagrangian. A mass-spring model with soil spring can be approximately constructed by solving the derived equations.

The applicability and limitation of the mass-spring model with a soil spring are readily understood following the procedure of formulation which includes two approximations. It should be emphasized that the mass-spring model with a soil spring solves the same physical problem of the continuum mechanics. The mass-spring model is a smart alternative to analyze a high-fidelity analysis model of a structure and soil with suitably prescribed boundary and interface conditions, to estimate the effects of the soil-structure interaction on the structural response.

Bibliography

Alhajyaseen, W. K. M., M. Asano and H. Nakamura, *et al.*, 2013. Stochastic approach for modeling the effects of intersection geometry on turning vehicle paths. *Transportation Research Part C: Emerging Technologies* 32:179–192.

Asai, M., Y. Miyagawa and I. N. Muhari, *et al.*, 2016. Coupled tsunami simulation based on a 2D shallow-water equation-based finite difference method and 3D incompressible smoothed particle hydrodynamics. *Journal of Earthquake and Tsunami* 10:5:1640019.

Baba, T., N. Takahashi and Y. Kaneda, *et al.*, 2014. Tsunami inundation modeling of the 2011 Tohoku Earthquake using three-dimensional building data for Sendai, Miyagi Prefecture, Japan. *Tsunami Events and Lessons Learned: Environmental and Societal Significance*, pp. 89–98.

Beeson, P., N. K. Jong and B. Kuipers, 2005. Towards autonomous topological place detection using the extended Voronoi graph. *Proceedings of the 2005 IEEE International Conference on Robotics and Automation*, pp. 4373–4379.

Bentley, J. L., 1975. Multidimensional binary search trees used for associative searching. *Communications of the ACM* 18:9:509–517.

Berenger, J., 1994. A perfectly matched layer for the absorption of electromagnetic waves. *Journal of Computational Physics* 114:185–200.

Boore, D. M., 1983. Stochastic simulation of high-frequency ground motions based on seismological models of the radiated spectra. *Bulletin of the Seismological Society of America* 73:1865–1894.

Cabinet Office, Japan, 2021a. *Tokyo Metropolis Earthquake.* `http://www.bousai.go.jp/jishin/syuto/index.html` (in Japanese, accessed September 30, 2021).

Cabinet Office, Japan, 2021b. *Nankai Trough Earthquake.* `http://www.bousai.go.jp/jishin/nankai/index.html` (in Japanese, accessed September 30, 2021).

Caendish, J. C., D. A. Field and W. H. Frey, 1985. An approach to automatic three-dimensional finite element mesh generation. *International Journal for Numerical Methods in Engineering* 21:329–347.

Campos, G. R. D., P. Falcone and J. Sjoberg, 2013. Autonomous cooperative driving: A velocity-based negotiation approach for intersection crossing. *16th International IEEE Conference on Intelligent Transportation Systems (ITSC 2013)*, pp. 1456–1461.

Cerjan, C., D. Kosloff and R. Kosloff, *et al.*, 1985. A nonreflecting boundary condition for discrete acoustic and elastic wave equations. *Geophysics* 50:705–708.

Clayton, R. and B. Engquist, 1977. Absorbing boundary conditions for acoustic and elastic wave equations. *Bulletin of the Seismological Society of America* 67:1529–1540.

Chen, J., T. Takeyama and H. O-Tani, *et al.*, 2019. Using high performance computing for liquefaction hazard assessment with statistical soil models. *International Journal of Computational Methods* 16:05:1840005.

Colombo, A. and D. Del Vecchio, 2015. Least restrictive supervisors for intersection collision avoidance: A scheduling approach. *IEEE Transactions on Automatic Control* 60:6:1515–1527.

Cristiani, E., B. Piccoli and A. Tosin, 2014. *Multiscale Modeling of Pedestrian Dynamics.* New York: Springer.

Curtis, S. and D. Manocha, 2014. Pedestrian simulation using geometric reasoning in velocity space. *Pedestrian and Evacuation Dynamics 2012*, pp. 875–890.

Dhingra, S. L. and I. Gull, 2008. *Traffic Flow Theory: Historical Research Perspectives.*

Dias, C., M. Iryo-Asano and T. Oguchi, 2017. Predicting optimal trajectory of left-turning vehicle at signalized intersection. *Transportation Research Procedia* 21:240–250.

Dijkstra, E. W., 1959. A note on two problems in connection with graphs. *Numerische Mathematik* 1:269–271.

Doniec, A., R. Mandiau and S. Piechowiak, *et al.*, 2008a. A behavioral multi-agent model for road traffic simulation. *Engineering Applications of Artificial Intelligence* 21:8:1443–1454.

Doniec, A., R. Mandiau and S. Piechowiak, *et al.*, 2008b. Anticipation based on constraint processing in a multi-agent context. *Autonomous Agents and Multi-Agent Systems* 17:2:339–361.

Fenves, G. L., D. P. Poland and R. T. Eguchi, *et al.*, 2011. *Grand Challenges in Earthquake Engineering Research: A Community Workshop Report.* National Academies Press.

Fu, Y., C. Li and B. Xia, *et al.*, 2016. A novel warning/avoidance algorithm for intersection collision based on dynamic Bayesian networks. *2016 IEEE International Conference on Communications (ICC)*, pp. 1–6.

Furuichi, M., D. Nishiura and M. Asai, *et al.*, 2017. The first real-scale DEM simulation of a sand-box experiment using 2.4 billion particles. *Proceedings of the International Conference for High Performance Computing, Networking, Storage and Analysis*, http://sc17.supercomputing.org/SC17%20Archive/tech_poster/tech_poster_pages/post113.html, Colorado, Denver, USA.

Gao, Y., 2011. Calibration of steady-state car-following models using macroscopic loop detector data. *75 Years of the Fundamental Diagram for Traffic Flow Theory. Greenshields Symposium* 149:178–198.

Geospatial Information Authority of Japan, 2019, *5m mesh digital elevation map (Tokyo ward area)*. http://www.gsi.go.jp/MAP/CD-ROM/dem5m/index.htm (in Japanese, accessed September 30, 2021).

Ghoshray, S. and K. K. Yen, 1996. A comprehensive robot collision avoidance scheme by two-dimensional geometric modeling. *Proceedings of IEEE International Conference on Robotics and Automation* 2:1087–1092.

Goto, Y., M. Affan and N. Agussabti, *et al.*, 2012. Tsunami evacuation simulation for disaster education and city planning. *Journal of Disaster Research* 7:92–101.

Gunawan, L. T., S. Fitrianie and C. K. Yang, *et al.*, 2012. TravelThrough: a participatory-based guidance system for traveling through disaster areas. *CHI '12 Extended Abstracts on Human Factors in Computing Systems*, pp. 241–250.

Helbing, D. and P. Molnar, 1995. Social force model for pedestrian dynamics. *Physical Review E*, 51:5:4282–4286.

Hisada Y., 1994. An efficient method for computing Green's functions for a layered half-space with sources and receivers at close depths. *Bulletin of the Seismological Society of America*, 84:1456–1472.

Hori, M., W. Lalith and T. Ichimura, *et al.*, 2014. Meta-modeling for constructing model consistent with continuum mechanics. *Journal of Japan Society of Civil Engineers*, A2:269–275.

Hori, M., 2015. Computational Earthquake Engineering. In *Encyclopedia of Earthquake Engineering*, ed. M. Beer, I. A. Kougioumtzoglou and E. Patelli, *et al.* New York: Springer-Verlag.

Hori, M., 2018. *Introduction to Computational Earthquake Engineering*, 3rd edition. Singapore: World Scientific.

Ichimura, T., M. Hori and J. Bielak, 2009. A hybrid multiresolution meshing technique for finite element three-dimensional earthquake ground Motion modeling in basins Including topography. *Geophysical Journal International* 177:1221–1232.

Ichimura, T., K. Fujita and S. Tanaka, *et al.*, 2014. Physics-based urban earthquake simulation enhanced by 10.7 BlnDOF × 30 K time-step unstructured FE non-linear seismic wave simulation. *SC'14 Proceedings of the International Conference for High Performance Computing, Networking, Storage, and Analysis*, pp. 15–26.

Ichimura, T., K. Fujita and P. E. B. Quinay, *et al.*, 2015. Implicit nonlinear wave simulation with 1.08T DOF and 0.270T unstructured finite elements to enhance comprehensive earthquake simulation. *SC'15 Proceedings of the International Conference for High Performance Computing, Networking, Storage, and Analysis*, Article No. 4.

Ichimura, T., R. Agata and T. Hori, *et al.*, 2016. An elastic/viscoelastic finite element analysis method for crustal deformation using a 3-D island-scale high-fidelity model. *Geophysical Journal International* 206:114–129.

Ichimura, T., K. Fujita and T. Yamaguchi, *et al.*, 2018. A fast scalable implicit solver for nonlinear time-evolution earthquake city problem on low-ordered unstructured finite elements with artificial intelligence and transprecision computing. *SC'18 Proceedings of the International Conference for High Performance Computing, Networking, Storage, and Analysis*, Article No. 49.

Ichimura, T., K. Fujita and M. Hori, *et al.*, 2020. A fast scalable iterative implicit solver with Green's function-based neural networks. *2020 IEEE/ACM 11th Workshop on Latest Advances in Scalable Algorithms for Large-Scale Systems (ScalA)*, pp. 61–68.

Idriss, I. M., R. D. Singh and R. Dobry, 1978. Nonlinear behavior of soft clays during cyclic loading. *Journal of the Geotechnical Engineering Division* 104:1427–1447.

Iiyama, K., A. Yoshiyuki and K. Fujita, *et al.*, 2019. A point-estimate based method for soil amplification estimation using high resolution model under uncertainty of stratum boundary geometry. *Soil Dynamics and Earthquake Engineering* 121:480–490.

Imamura, F., A. Muhari and E. Mas, *et al.*, 2012. Tsunami disaster mitigation by integrating comprehensive countermeasures in Padang city, Indonesia. *Journal of Disaster Research* 7:1:48–64.

Irikura, K., 1986. Prediction of strong acceleration motions using empirical Green's function. *Proceedings of the 7th Japan Earthquake Engineering Symposium*, pp. 151–156.

Isler, V., D. Sun and S. Sastry, 2005. Roadmap based pursuit-evasion and collision avoidance. *Proceedings of Robotics: Science and Systems*, pp. 257–264.

Javier, A. M., N. Tobias and S. Roland, *et al.*, 2015. Collision avoidance for aerial vehicles in multi-agent scenarios. *Autonomous Robots* 39:1:101–121.

Japan Meteorological Agency, 1995. *Strong ground motion of The Southern Hyogo prefecture earthquake in 1995 observed at Kobe JMA observatory.* `http://www.data.jma.go.jp/svd/eqev/data/kyoshin/jishin/hyogo_nanbu/dat/H1171931.csv` (in Japanese, accessed September 30, 2021).

Jiang, R. and Q. S. Wu, 2006. The moving behavior of a large object in the crowds in a narrow channel. *Physica A: Statistical Mechanics and its Applications* 364:457–463.

Jiang, S., J. Ferreira, Jr. and M. C. Gonzalez, 2017. Activity-based human mobility patterns inferred from mobile phone data: a case study of Singapore. *IEEE Transactions on Big Data* 3:2:208–219.

Kuligowski, E. D. and R. D. Peacock, 2005. A review of building evacuation models. *Technical Note* 1471, National Institute of Standards and Technology, U.S. Dept. of Commerce.

Kwon, Y. I., S. Morichi and T. Yai, 1998. Analysis of pedestrian behavior and planning guidelines with mixed traffic for narrow urban streets. *Transportation Research Record* 1636:1:116–123.

Leonel, A. E .M., M. L. L. Wijerathne and M. Hori, *et al.*, 2013. On the development of an MAS based evacuation simulation system: autonomous navigation and collision avoidance. *Lecture Notes in Computer Science* 8291:388–395.

Leonel, A. E. M., M. L. L. Wijerathne and M. Hori, *et al.*, 2013. On the development of an MAS based evacuation simulation system: autonomous navigation and collision avoidance. *Lecture Notes in Computer Science* 8291:388–395.

Leonel, A. E. M., M. L. L. Wijerathne and M. Hori, *et al.*, 2014b. A scalable workbench for large urban area simulations, comprised of resources for behavioural models, interactions and dynamic environments. *Procedia Computer Science* 108:937–947.

Leonel, A. E. M., M. L. L. Wijerathne and M. Hori, *et al.*, 2016. Automatic evacuation management using a Multi Agent System and parallel metaheuristic search. *Lecture Notes in Computer Science* 9862:387–396.

Leonel, A. E. M., M. L. L. Maddegedara and T. Ichimura, *et al.*, 2017. On the performance and scalability of an HPC enhanced Multi Agent System based evacuation simulator. *Procedia Computer Science* 108:937–947.

Leonel, A. E. M., M. L. L. Wijerathne and S. Jacob, *et al.*, 2019. Mass evacuation simulation considering detailed models: behavioral profiles, environmental effects, and mixed-mode evacuation. *Asia Pacific Management Review* 24:2:114–123.

Lo, S. H., 1985. A new mesh generation scheme for arbitrary planar domains. *International Journal for Numerical Methods in Engineering* 21:1403–1426.

Lu, X., B. Hana and M. Hori, *et al.*, 2014. A coarse-grained parallel approach for seismic damage simulations. *Advances in Engineering Software* 70:90–103.

Lu, X. and H. Guan, 2017. *Earthquake Disaster Simulation of Civil Infrastructures – From Tall Buildings to Urban Areas.* Singapore: Springer Singapore.

Makinoshima, F., F. Imamura and Y. Abe, 2018. Enhancing a tsunami evacuation simulation for a multi-scenario analysis using parallel computing. *Simulation Modelling Practice and Theory* 83:36–50.

Mas, E., S. Koshimura and F. Imamura, *et al.*, 2015. Recent advances in agent-based tsunami evacuation simulations: case studies in Indonesia, Thailand, Japan and Peru. *Pure and Applied Geophysics* 172:3409–3424.

Masing, G., 1926. Eigenspannungen und verfestigung beim messing. *Proceedings of the 2nd International Congress of Applied Mechanics*, pp. 332–335 (in Germany).

Miyamura, T., H. Akiba and M. Hori, 2015. Large-scale seismic response analysis of super-high-rise steel building considering soil-structure interaction using K computer. *International Journal of High-Rise Buildings* 4:1:75–83.

Miyamura, T., S. Tanaka and M. Hori, 2016. Large-scale seismic response analysis of a super-high-rise-building fully considering the soil-structure interaction using a high-fidelity 3D solid element model. *Journal of Earthquake and Tsunami* 10:5.

Miyazaki, H., Y. Kusano and N. Shinjou, *et al.*, 2012. Overview of the K computer system. *Fujitsu Scientific and Technical Journal* 48:302–309.

Mori, M. and H. Tsukaguchi, 1987. A new method for evaluation of level of service in pedestrian facilities. *Transportation Research Part A: General* 21:3:223–234.

National Research Institute for Earth Science and Disaster Prevention, 2021. *Geo-Station*. https://www.geo-stn.bosai.go.jp/ (in Japanese, accessed September 30, 2021).

National Research Institute for Earth Science and Disaster Prevention, 2021. *Strong-motion seismograph networks (K-NET, KiK-net)*. http://www.kyoshin.bosai.go.jp/ (accessed September 30, 2021).

O-Tani, H., J. Chen and M. Hori, 2014a. Template-based floor shape recognition applied to 3D building shapes of GIS data. *Journal of Japan Society of Civil Engineers* 70:4:1124–1131.

O-Tani, H., J. Chen and M. Hori, 2014b. Automatic combination of the 3D shapes and the attributes of buildings in different GIS data. *Journal of Japan Society of Civil Engineers* 70:2:631–639.

Ouellette, M. J. and M. S. Rea, 1989. Illuminance requirements for emergency lighting. *Journal of the Illuminating Engineering Society* 18:1:37–42.

Papadrakakis, M., P. Vagelis and N. D. Lagaros, 2017. *Computational Methods in Earthquake Engineering*, Vol. 3. New York: Springer-Verlag.

Piegl, L. and W. Tiller, 1997. *The NURBS Book*, 2nd edition. New York: Springer-Verlag.

Pournaras, E., J. Nikolic and P. Velasquez, *et al.*, 2016. Self-regulatory Information Sharing in Participatory Social Sensing. *The European Physical Journal Data Science* 5:14.

Poursartip, B., A. Fathi and J. L. Tassoulas, 2020. Large-scale simulation of seismic wave motion: A review. *Soil Dynamics and Earthquake Engineering* 129:105909.

Rene, M., C. Alexis and A. Jean-Michel, *et al.*, 2008. Behaviour based on decision matrices for a coordination between agents in a urban traffic simulation. *Applied Intelligence* 28:2:121–138.

Riaz, M. R., H. Motoyama and M. Hori, 2021. Review of soil-structure interaction based on continuum mechanics: theory and use of high performance computing. *Geosciences* 11:2:72.

Saad, Y., 2003. *Iterative Methods for Sparse Linear Systems*. SIAM.

Sahil, N., B. Andrew and C. Sean, *et al.*, 2015. Generating pedestrian trajectories consistent with the fundamental diagram based on physiological and psychological factors. *PLOS One* 10:4:1–17.

Sahin, A., R. Sisman and A. Askan, *et al.*, 2016. Development of integrated earthquake simulation system for Istanbul. *Earth Planets Space* 68:115–135.

System for Integrated Simulation of Earthquake and Tsunami Hazard and Disaster, Research and Application Development in Post-K Project for Social and Scientific Important Tasks, Research Project of Disaster Prevention and Environment Problem, 2020. *System for Integrated Simulation of Earthquake and Tsunami Hazard and Disaster.* `https://www.eri.u-tokyo.ac.jp/LsETD/Post_K/bosaitop/eng/index.html` (accessed September 30, 2021).

Tanaka, S., M. Hori and T. Ichimura, 2016. Hybrid finite element modeling for seismic structural response analysis of a reinforced concrete structure. *Journal of Earthquake and Tsunami* 10:1640015.

Theodore, V. G. and B. Ellingwood, 1986. Serviceability limit states: deflection. *Journal of Structural Engineering* 112:1:67–84.

van den Berg J., S. J. Guy and M. Lin, *et al.*, 2011. Reciprocal n-body collision avoidance. *Robotics Research* 70:3–9.

Wang, Z. and G. Jia, 2021. A novel agent-based model for tsunami evacuation simulation and risk assessment. *Natural Hazards* 105:2045–2071.

Weidmann, U., 1993. Transporttechnik der Fussganger, *Transporttechnische Eigenschaften des Fussgangerverkehrs (Literturauswertung)* 90 (in Germany).

Wijerathne M. L. L., L. A. Melgar and M. Hori, *et al.*, 2013. HPC enhanced large urban area evacuation simulations with vision based autonomously navigating Multi Agents. *Procedia Computer Science* 18:1515–1524.

Wijerathne M. L. L., P. Wasuwat, A. Leonel, *et al.*, 2018. Scalable HPC enhanced agent based system for simulating mixed mode evacuation of large urban areas. *Transportation Research Procedia* 34:275–282.

Winget, J. M. and T. J. R. Hughes, 1985. Solution algorithms for nonlinear transient heat conduction analysis employing element-by-element iterative strategies. *Computer Methods in Applied Mechanics and Engineering* 52:711–815.

Xin, Z. and C. Gang-Len, 2014. A mixed-flow simulation model for congested intersections with high pedestrian vehicle traffic flows. *SIMULATION* 90:5:570–590.

Xue, M. and O. Ryuzo, 2012. Examination of vulnerability of various residential areas in China for earthquake disaster mitigation. *Procedia – Social and Behavioral Sciences* 35:369–377.

Yokomatsu, M., Y. Akiyama and Y. Ogawa, *et al.*, 2017. Numerical analysis of dynamic planning problems on corporate recovery capital investment considering various disaster scenarios. *Journal of Disaster Research* 14:3:508–520.

Yoshimura, S., M. Hori, and M. Ohsaki, 2015. *High-Performance Computing for Structural Mechanics and Earthquake/Tsunami Engineering*. New York: Springer International Publishing.

Young-In, K., S. Morichi and T. Yai, 1998. Analysis of pedestrian behavior and planning guidelines with mixed traffic for narrow urban streets. *Transportation Research Record: Journal of the Transportation Research Board* 1636:116–123.

Zienkiewicz, O. C. and R. L. Taylor, 2005. *The Finite Element Method for Solid and Structural Mechanics*. New York: Elsevier.

Index